THE CHEMISTRY OF COLA

First American Edition 2020
Kane Miller, A Division of EDC Publishing

Copyright © UniPress Books Ltd 2020
Published by arrangement with UniPress Books Ltd.
All rights reserved. No part of this book may be reproduced, transmitted
or stored in an information retrieval system in any form or by any means,
graphic, electronic or mechanical, including photocopying, taping and recording,
without prior written permission from the publisher.

For information contact:
Kane Miller, A Division of EDC Publishing
PO Box 470663
Tulsa, OK 74147-0663
www.kanemiller.com
www.edcpub.com
www.usbornebooksandmore.com

Library of Congress Control Number: 2019936214
Printed in China
ISBN: 978-1-68464-003-4

1 2 3 4 5 6 7 8 9 10

SAFETY NOTICE: The experiments in this book must be conducted under adult
supervision, and with all reasonable caution including an awareness of food allergies and
intolerance. The instructions provided in each experiment are no replacement for the
sound judgment of participants. The author and publisher accept no liability for any
mishap or injury arising from participation in the experiments.

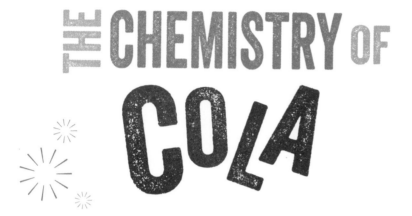

THE CHEMISTRY OF COLA

THE CURIOUS WORLD OF KITCHEN SCIENCE

DR. JAMIE GALLAGHER

Kane Miller
A DIVISION OF EDC PUBLISHING

CONTENTS

CHAPTER 4: FOOD CHEMISTRY

CHAPTER 5: LOOKING CLOSELY AT COLA

CHAPTER 6: COLA ON THE GO

INTRODUCTION

Chemistry is all around us—it is the medicine we take, the air that we breathe, and the cola that we drink. Chemistry is the branch of science that explores matter, from atoms (the small building blocks of matter) through to large, complex molecules made of thousands of atoms connected together. Chemists try and understand what things are made of, how they are made, and why they behave the way they do.

Chemistry is sometimes called the central science because it connects and overlaps with so many other areas of science. Physics tells us how the planets move, but chemistry tells us what they are made of and if they are habitable. Biology can describe the process of natural selection, while chemistry lets us understand the bonding of our DNA.

With this book as your companion you will see the world around you in a whole new light. Using everyday examples, you will discover some of the most important and interesting aspects of chemistry. As you go through the chapters of this book, your kitchen will become your laboratory as you are guided step-by-step through experiments, most requiring nothing more than some common household items and ingredients.

Why are sugary drinks so bad for your teeth? Find out on page 56.

Ice and fire, water and oil, fruits and flavors—as you read through these pages you will explore a huge number of topics, from the food we eat to the trash we throw away. You will discover what acids are and how to test for them at home. You will shrink down to explore the worlds of atoms and molecules. You will grow crystals, melt plastic, create clouds, and crush bottles—all in the name of chemistry.

Experiment with chromatography on pages 118-119!

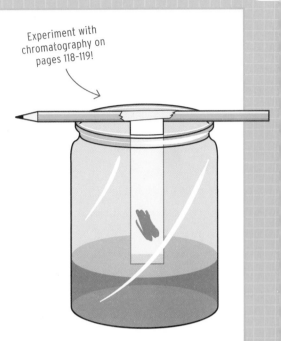

The science you will uncover in this book will help you understand chemistry, whether that is lab-based experiments or the large-scale reactions that take place in the environment. Along the way you will get lots of opportunity to test your newfound chemistry knowledge, with quizzes and questions throughout the book.

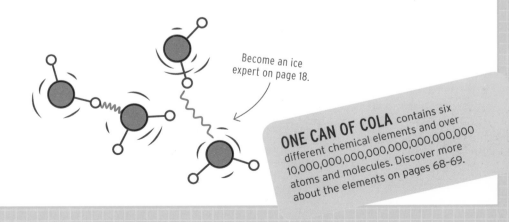

Become an ice expert on page 18.

ONE CAN OF COLA contains six different chemical elements and over 10,000,000,000,000,000,000,000 atoms and molecules. Discover more about the elements on pages 68-69.

CHAPTER 1
STATES OF MATTER

DISCOVER...

LEARN...

EXPERIMENT...

DISCOVER: THE CHEMISTRY OF COLA

The chemistry of cola is about a lot more than taste. Cola is instantly recognizable due to its distinctive color, and that's no accident. Food scientists and chemists have long worked with marketing specialists to come up with products that are a feast for the eyes as well as the mouth. This section takes a closer look at how cola gets its color.

Imagine you decided to make your own cola using the ingredients listed on the bottle. Mix carbonated water, sugar, phosphoric acid, and caffeine, and you will be left with a clear, colorless fizzy liquid. Even adding the flavorings won't change the appearance much. It's not until you add E150d that cola looks like cola. But why do the manufacturers bother to add the color at all? It's all to do with cola's history.

CREATING THE COLOR

Since its creation in 1886, manufacturers have worked to preserve cola's distinctive taste and appearance, despite changes in the recipe. The original recipe has become the stuff of legend—it was reportedly shared with only four people before the death of its creator, Dr. Pemberton, in 1888. Most of the recipe has been pieced together over the years, and it's known that the blend of flavors comes from various oils (citrus, cinnamon, nutmeg, and more), and that the bitter kola nut was used to add caffeine. To mask the bitterness of the kola nut, Pemberton added sugar to make the drink sweet—some sugar he dissolved, and the rest he first turned into caramel. It was the adding of the caramel that gave his cola its distinctive color.

Dried kola nuts

Although caramel is no longer added to cola, manufacturers retain the appearance of the drink by adding caramel colors. For that purpose, E150d—a soluble coloring agent which can range from yellow to dark brown—is used. You'll find this color added to many yellow or brown foods and drinks, from coffee to pet food. Adding this color contributes very little to the taste of cola, but it keeps it looking the way customers expect.

SEE pages 122–123 to find out what happens when you mix cola with milk!

Chemists can change the color of foods using E numbers 100–199, even making silver and gold foods. People often worry about additives in food, but seeing these numerical codes listed on a label isn't a cause for alarm. The codes are for convenience and satisfy legal requirements. The numbers refer to specific chemicals—whereas "brown food coloring" might be considered vague, "E150d" is specific. The codes aren't just used for colors; E300 is the code for vitamin C.

TRICKING THE EYE

Adding color to drinks to make them more appealing isn't the only trick in the chemist's book. The next time you go to the supermarket, take a close look at the bottled water. What color are the bottles? You might think they are all clear, just like cola bottles, but some of the pricier bottles are actually slightly blue. A pigment is added to the plastic bottles during manufacture because people tend to view slightly blue bottles as cleaner and fresher than transparent bottles.

DISCOVER: SOLID, LIQUID, GAS

There are lots of ways we can begin to understand the world around us, and the first step for a chemist is often to think about solids, liquids, and gases—the states of matter. With a bottle of cola, we have a solid bottle that holds a liquid, and that liquid holds a gas.

When cola is made, carbon dioxide (CO_2) gas is dissolved in the liquid. The liquid is then kept at high pressure to stop the gas from escaping—until the bottle is opened and you hear that distinctive hiss. When the bottle is open, the CO_2 escapes by forming bubbles, which rise to the surface and pop in the atmosphere. This is also why you burp after drinking cola: the low pressure in your stomach allows the CO_2 to escape from the liquid.

Understanding the states of matter and how substances behave at the particle level (the level of atoms or molecules) is important in helping chemists and physicists understand the world around them.

SOLIDS

In solids, the atoms or molecules (see pages 66–67 and 78–79) touch their neighbors, meaning they are trapped. This is why a solid has a fixed volume and shape. But even though the particles are trapped, they still vibrate a small amount.

LIQUIDS

In a liquid, the particles have more freedom to move; they can move past each other, and move around randomly. We can see the result of this microscopic particle movement: liquids will take the shape of the container and can be poured.

GASES

In a gas, the particles can be very far apart. They don't cling to their neighbor, as in a solid, or flow past each other, as with a liquid. They bounce around randomly, colliding with each other and with container walls. Gases have a large volume and low density, compared with liquids and solids at the same temperature, making them very light. This is why the bubbles of gas float to the top of your cola.

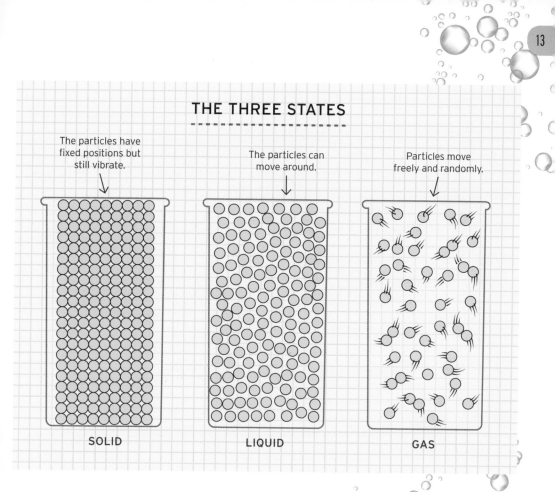

THE THREE STATES

The particles have fixed positions but still vibrate.

The particles can move around.

Particles move freely and randomly.

SOLID

LIQUID

GAS

CHANGING PHASE

A substance can be transformed from solid to liquid, and liquid to gas, by heating. This doesn't mean that gases are always hotter than liquids —different substances change phase at different temperatures. At room temperature, carbon dioxide is a gas, but water (H_2O) is a liquid. The size, shape, and charge of the particles in a substance determine what state it will be in at any given temperature.

SUBLIME MATERIALS

Some substances can move from solid straight to gas, a process that is called sublimation. Water can sublime; you might notice that ice cubes slowly (very slowly) shrink in your freezer. This is because some of the water particles gradually escape as a gas.

DISCOVER: WHEN A LIQUID ISN'T A LIQUID

The previous section looked at solids, liquids, and gases, but that's only part of the story! Some substances can behave very strangely. Sometimes just looking at something, or touching it, isn't enough to decide if it is a solid, liquid, or gas.

Although plastic bottles look and feel like solids, they are made from a very complex material. Plastics are made of very large particles—long chains that can move past each other like a liquid, yet because they are so long, they get tangled up. It's like when you put your earphones in your pocket and they end up in a knotted ball. The particles in a plastic bottle get so twisted that they hold a solid form, and yet some of them are still able to move and flow.

Another example of a liquid disguising itself as a solid is asphalt, which is used to surface roads. In 1927, Australian professor Thomas Parnell put some asphalt in a funnel and . . . waited. Gradually, the "solid" asphalt flowed down the funnel, with each drop taking a very long time to fall. The experiment has been running for over 90 years, and there have been only nine drops of this very thick liquid!

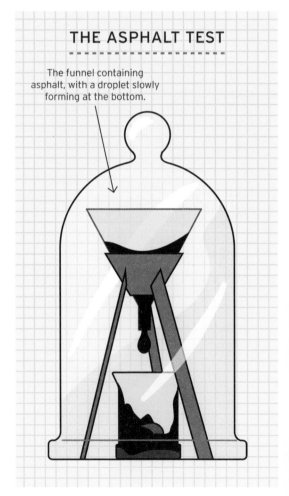

THE ASPHALT TEST

The funnel containing asphalt, with a droplet slowly forming at the bottom.

NON-NEWTONIAN FLUIDS

Some liquids will change depending on how you treat them. These are called non-Newtonian fluids. How solid or liquid so-called non-Newtonian fluids are depends on the force applied to them. Some liquids can act like solids if you hit them hard, and others might be made even runnier.

Ketchup is a non-Newtonian fluid; it can be made runnier by applying force. You might have discovered this if you've tried to put ketchup on your hot dog, but nothing came out, only to find it splattered everywhere when you shook the bottle. When sitting in the bottle, ketchup is quite thick, but if you shake it, the thick liquid momentarily becomes thin and runs out of the bottle.

Engineers at the University of Melbourne conducted a study of how best to get ketchup out of a bottle. They discovered it's best to hold the bottle over your food at a 45-degree angle, then tap the bottle gently, increasing the force of the taps until the sauce flows. Many manufacturers are now getting around the messy ketchup problem by making the bottles plastic and squeezable.

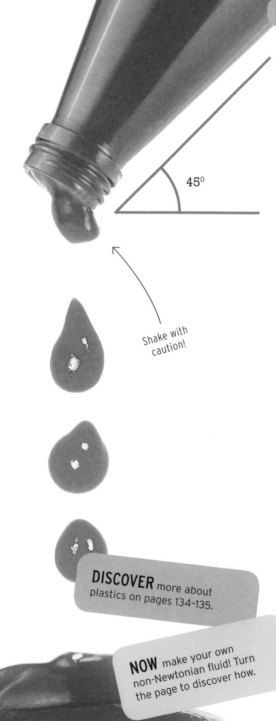

45°

Shake with caution!

DISCOVER more about plastics on pages 134-135.

NOW make your own non-Newtonian fluid! Turn the page to discover how.

EXPERIMENT: NON-NEWTONIAN FLUID

The previous pages explained that some liquids behave differently, depending on how you treat them. Now it's time to put this to the test. In this messy experiment, you will make a non-Newtonian fluid (slime!), and see how it reacts in different ways.

YOU WILL NEED:

- Large mixing bowl
- 300–500 g (10–17 oz.) of cornstarch
- 400–600 ml (13–20 fl. oz.) of water

WHAT TO DO:

1. Make sure you have a clear, flat surface to work on—this can get messy! Put some old newspapers, magazines, or plastic sheeting over your work area to make cleaning up easier.

2. Put the cornstarch in the bowl (the more cornstarch, the more slime). Keep 75 g (2½ oz.) in reserve in case you need to add to the mixture later. Use your hands to break up any large clumps.

3. Start adding water slowly. You will need about 75 ml (2½ fl. oz.) for every 100 g (3½ oz.) of cornstarch you use. Add 200 ml (7 fl. oz.) of water and mix with your hands. Continue slowly adding water and mixing until it has a thick yet still runny consistency.

4. If the mixture is too runny, add some of the reserve cornstarch.

TIME TO TEST

Try striking the mixture with short, sharp punches, withdrawing your hand quickly. It should feel like you are punching a solid, and there shouldn't be any splashes! Next, put your hand on top of the mixture and let your hand slowly sink into it.

When force is applied rapidly to the cornstarch mixture, it behaves like a solid, but without a sudden force the solution remains a liquid. You might even be able to scoop up some of the mixture into your hand and roll it quickly into a ball. As soon as you stop rolling, the ball will revert to behaving like a liquid.

WHAT HAPPENS?

The solution you have made is a thick suspension of large cornstarch particles. When the liquid moves slowly, the particles can slide past each other, but when the mixture is hit quickly, the particles can't move out of the way in time.

Imagine you want to get to a restroom on the other side of a crowd of people: if you run as fast as you can, you'll just crash into people and not get very far. If, however, you slowly work your way through, people will shuffle aside, allowing you to pass.

REMEMBER: Once you are finished with the solution, dispose of it as food waste. Cornstarch can clog drains.

DISCOVER: ICE SURPRISE

On a hot summer day, an ice-cold soft drink can be very refreshing. You drop the cubes into the glass, hear the distinctive fizz as they produce lots of bubbles, and then there is the satisfying crack of the ice. For a chemist, this isn't just refreshing, it's fascinating—ice is an amazing and surprising material.

One of the most amazing things about ice is that it floats! This might not seem very interesting at first glance, but ice is one of a few special materials that have a solid form less dense than their liquid form. Pages 12–13 showed that solids are closely packed materials, while liquids have particles that can move around more freely. In almost every case, the liquid has more space between the particles, meaning the liquid is less dense and would float on the solid. Ice is a special exception.

To understand what is happening to make the ice less dense, we need to think about the atoms in the material and how they arrange into molecules. Water (H_2O) has a central oxygen atom with two hydrogen atoms attached. Hydrogen atoms and oxygen atoms are quite different; when they are bonded, the molecules

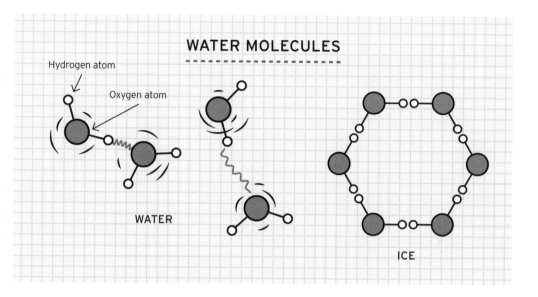

WATER MOLECULES

Hydrogen atom

Oxygen atom

WATER

ICE

bonded, the molecules end up with tiny charges. The hydrogen atoms get tiny positive charges, and the oxygen atoms get tiny negative charges. As negative and positive charges attract each other, the hydrogens of one H_2O molecule will be attracted to the oxygen of another H_2O molecule. These attractions are called hydrogen bonds. At room temperature, the molecules vibrate around so much with thermal energy that these weak hydrogen bonds don't last long before the molecules drift away from each other. This all changes at 4°C (39°F).

HEXAGONAL PATTERNS

As water gets colder, the molecules wiggle less and less, and the hydrogen bonds last a little longer. At 4°C (39°F), the hydrogen bonds start to become stronger than the thermal wiggles and the molecules start to bond to each other in a regular way. The water molecule is shaped like a "V" and the hydrogen bonds allow the water molecules to form hexagonal patterns which have bigger gaps between the molecules than in liquid water. At 0°C (32°F), a solid has formed that is 8.3 percent less dense than liquid water.

It's for this reason that ice floats on water. It also stops oceans from freezing from the bottom up, and allows icebergs to float on an ocean's surface, keeping sea levels down. And of course, ice allows us to have a glass of refreshing cola on a hot day.

LEARN ABOUT: KITCHEN CHEMISTRY

Try this quick-fire quiz to test your general chemistry knowledge. You might need to read ahead in the book for help.

POP QUIZ: IN THE KITCHEN

1. The bubbles in soft drinks are composed of which gas?
a) Oxygen
b) Nitrogen
c) Carbon dioxide
d) Dihydrogen monoxide

2. Baking powder is a mix of a carbonate, an acid, and a filler, which is added to bread to make it rise. What is happening in this important chemical process? (Look ahead to page 46 for more information.)
a) The baking powder reacts with sugars in the bread to produce carbon dioxide.
b) The acidic baking powder reacts with the neutral water, producing oxygen bubbles.
c) Baking powder expands rapidly due to heat, then on cooling to room temperature it rapidly contracts, leaving pockets of air behind.
d) The dried ingredients in baking powder begin to react when exposed to water, producing carbon dioxide.

3. In cooking, the Maillard reaction is incredibly important. But what is it? (Turn to page 31 for more information.)
a) A reaction between sugars and amino acids that causes browning
b) A breakdown of potentially harmful bacteria
c) Evaporation of water, which creates a crisp surface
d) A softening of proteins, which makes meat tender

4. Caffeine is found in tea, coffee, and cola. It is also found in over 60 varieties of plants. Why is caffeine present in plants? (See page 110.)
a) It speeds up growth.
b) It is a natural predator deterrent.
c) It strengthens cell walls.
d) It is a by-product of pollen production.

5. Microwaves can be used to heat food very quickly, but how do they do this? (See pages 30-31.)
a) Air molecules trapped in food absorb microwaves and move rapidly.
b) The microwaves break the bonds in the food, causing energy to be released.
c) Through the ionization (the creation of electrical charge) of particles.
d) They cause the water molecules to spin rapidly, which causes heating.

LEARN ABOUT: SUBSTANCES AND STATES

Test your knowledge of substances and their states with this challenge. Match up the terms on the left with the correct definition on the right.

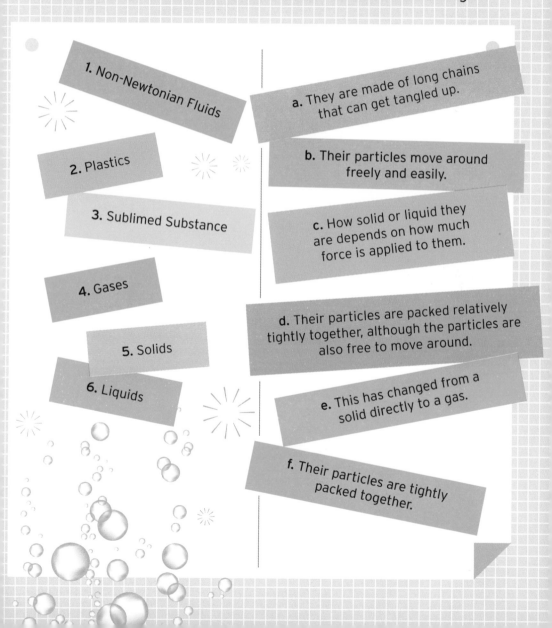

1. Non-Newtonian Fluids

2. Plastics

3. Sublimed Substance

4. Gases

5. Solids

6. Liquids

a. They are made of long chains that can get tangled up.

b. Their particles move around freely and easily.

c. How solid or liquid they are depends on how much force is applied to them.

d. Their particles are packed relatively tightly together, although the particles are also free to move around.

e. This has changed from a solid directly to a gas.

f. Their particles are tightly packed together.

EXPERIMENT: EXPANDING ICE

In this experiment you will measure the density change when water freezes into ice. For this, you will first need to freeze water and measure the volume change.

YOU WILL NEED:

- 500-ml (1-pt.) plastic measuring pitcher
- 250-ml (8-fl. oz.) measuring pitcher
- Water
- Freezer

WHAT TO DO:

1. Add 350 ml (12 fl. oz.) of room-temperature water to the large measuring pitcher.

2. Place the pitcher carefully in the freezer, making sure that it sits flat. (Ask an adult for help.)

3. Wait until the water is completely frozen before removing the pitcher from the freezer. You need to work out the new volume, but the surface might be irregular, so an extra step is needed.

4. Leave the pitcher to stand for 5–10 minutes—this will avoid thermal shock (which could break the pitcher) when extra water is added.

5. Add cold water on top of the ice and fill up to the 500-ml mark. Pour the water you just added into the small measuring pitcher and record the volume.

6. The volume of the ice is equal to 500 ml, minus the extra water you added.

ADULT SUPERVISION REQUIRED

WARNING: Glass measuring pitchers can shatter if exposed to sudden temperature changes. It is recommended that you use a strong plastic pitcher for the freezing.

MEASURING DENSITY

Density (ρ) is defined as the mass (m) divided by the volume (V). This is written as:

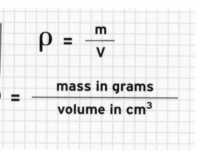

$$\rho = \frac{m}{V}$$

$$\rho = \frac{\text{mass in grams}}{\text{volume in cm}^3}$$

You have measured the volume, and because you are using water, you already know its mass. One milliliter ($\frac{1}{5}$ teaspoon) of water has a mass of one gram ($\frac{1}{25}$ oz.) at room temperature, so if you have 350 ml, it will weigh 350 g.

To calculate the above formula, you need to convert milliliters into cubic centimeters (cm^3). Luckily, 1 ml is equal to 1 cm^3, so you can use the numbers you measured.

To calculate the density of water, you must complete the formula:

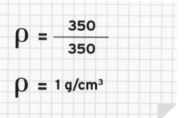

$$\rho = \frac{350}{350}$$

$$\rho = 1 \text{ g/cm}^3$$

Let's calculate the density of the ice (unless you spilled some, the mass should be the same as the water).

To calculate the change in density, use the formula:

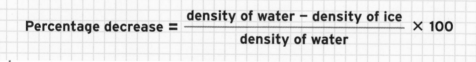

$$\text{Percentage decrease} = \frac{\text{density of water} - \text{density of ice}}{\text{density of water}} \times 100$$

Ice is reported to be 8.3% less dense than water. How close did you get? Where might any errors or uncertainty lie in your experiment?

EXPERIMENT: SUPERCOOL ICE

This is a tricky but very satisfying experiment. When it works, you are able to freeze a bottle of water instantly, before your eyes! To do this, you need to get water below its freezing point—without it actually freezing.

Once ice starts to appear, it can grow rapidly in a cold, wet environment. But forming the first few crystals can be challenging. Ice needs some assistance to start forming. Whether it's a small particle or a rough surface, water requires a solid base to begin freezing.

YOU WILL NEED:

- 3 sealed half-liter (16.9 oz.) plastic bottles of water (the lower the mineral content, the better)
- Hammer
- Heavy plastic or canvas bag
- 300 g (11 oz.) of salt
- Bucket
- 3 kg (7 lb.) of ice cubes
- Wooden stick
- Thermometer (optional, though recommended)

ADULT SUPERVISION REQUIRED

STEP 2   STEP 3   STEP 4

WHAT TO DO:

1. Put the ice cubes in the heavy plastic or canvas bag and break them up with the hammer. You should do this outside, on solid ground. Ask an adult for help.

2. Add the ice cubes to the bucket and then add some tap water, until the water reaches to half the level of the ice.

3. Sprinkle the 300 g of salt into the bucket and mix, using the stick (the water will be very cold).

4. Put the bottles of water into the bucket, so they are surrounded by ice.

5. Place the thermometer (if you have one) at a similar depth and position like the bottles.

6. The bottles must be kept very still from now on.

7. Wait 30–45 minutes, or until the thermometer reads below −6°C (21°F).

8. Carefully and slowly remove one of the bottles. Hopefully the water will still be liquid, despite it being below 0°C.

9. Bang the bottle down on a table and watch what happens. If you have managed to supercool the water, the bottle will instantly freeze! If it doesn't work, wait another 10 minutes, then try with the second bottle.

WHAT HAPPENS?

The sudden force of the bottle hitting the table is enough to allow some particles of ice to form. And as soon as some form, the whole bottle will freeze in a matter of seconds.

Once you get the hang of it, try gently opening a supercooled bottle of water and pouring it into a glass filled with ice cubes, and watch what happens.

COLD CLOUDS

Up in the atmosphere, ice often also needs assistance in order to form. High-altitude clouds can contain water droplets that can reach -30°C (-22°F) before they freeze. However, the droplets will freeze at much higher temperatures if there is dust in the air—the particles help to start the formation of the ice.

EXPERIMENT: CLOUD IN A JAR

Clouds are collections of water droplets that are derived from the gaseous water vapor produced when sunlight heats ocean water. The water vapor cools as it is driven by the wind up into the atmosphere. It then condenses onto the numerous particles of dust, smoke, and salt that fill the sky, forming water droplets.

The wind keeps these collections of water droplets up in the atmosphere as clouds. As more water vapor is driven up from the ocean to condense onto existing droplets, the clouds get bigger, until eventually the water droplets become too big and heavy to stay in the air. This causes droplets to fall to the ground. As the droplets descend they merge with other water droplets to become even larger, eventually emerging as rain. The amount of rain a cloud can produce depends on its size and the concentration of water inside it.

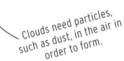

Clouds need particles, such as dust, in the air in order to form.

Dark storm clouds contain a higher concentration of water than white fluffy clouds do, which is why they can produce so much rain. A single storm cloud can release over 500 million liters (132 million gallons) of water.

Clouds usually form far out at sea, before being blown over land. But we can also watch this process happen on a much smaller scale in a jar.

YOU WILL NEED:

- Small glass jar with a lid
- Hot water
- Ice cubes
- Hair spray

ADULT SUPERVISION REQUIRED

WHAT TO DO:

1. Unscrew the lid from the jar, turn the lid upside down and place three or four ice cubes onto it.

2. Fill the jar half full with freshly boiled water. Ask an adult for help.

3. Balance the upturned lid on top of the jar.

4. Wait for five minutes. What do you see?

5. Empty the jar and repeat the experiment, but this time add a squirt of hair spray to the jar after half filling it with freshly boiled water but before balancing the upturned lid on top.

6. Wait for five minutes. What do you see this time?

WHAT HAPPENS?

In the first half of the experiment, though tiny droplets of water formed on the underside of the lid as the water vapor emitted by the hot water cooled and condensed, you won't have seen clouds. In the second half of the experiment, however, you should have seen a cloud form at the top of the jar. The hair spray performed the same function as dust in the atmosphere, providing particles that the water vapor could condense onto as it was cooled by the cold lid, thereby producing water droplets. If you removed the lid, you should have seen your cloud float away.

EXPERIMENT: ICE AND OIL

When you add ice to your drink, the ice cubes will float and gradually melt away. But since cola is almost entirely water, it isn't possible to see the melted ice mix with the liquid already in the glass. So, in this experiment, you will see very clearly the density difference between water and ice by separating the two with a layer of oil.

Water and oil don't like to mix. Pages 18–19 explained that water is made of one oxygen and two hydrogen atoms in a "V" shape. This molecule is "polar," which means it has a negative charge at one side and a slight positive charge at the other. Oils, on the other hand, are made of long carbon chains and are nonpolar (or, less polar than water).

Polar molecules are drawn to other polar molecules, so water molecules like to stick together. If you try and mix a nonpolar liquid with water, the water molecules will draw together because of their uneven electron charge and be repelled by the nonpolar liquid.

Since the water and oil won't mix, the least dense liquid will form a layer on top of the other one. Water is denser than most cooking oils, so in this experiment the vegetable oil will float on the surface of the water. Ice, however, is less dense than water, and also less dense than many cooking oils, so the oil will float on the water, and the ice will float on the oil.

LONG-CHAIN OIL MOLECULE

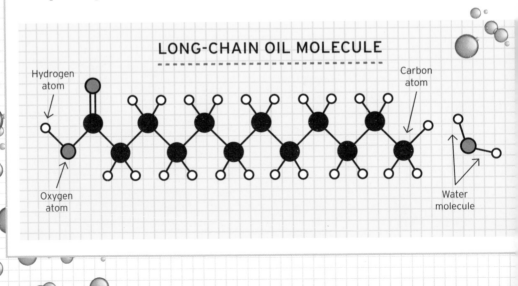

Hydrogen atom

Carbon atom

Oxygen atom

Water molecule

YOU WILL NEED:

- Tall, thin, transparent glass, tube, or measuring cylinder
- 200 ml (7 fl. oz.) of vegetable oil
- 200 ml of water
- Red or blue food coloring (optional)
- Ice cube tray (and water to fill it)

WHAT TO DO:

1. Make ice cubes (ice cubes with food coloring will be easier to see, but this isn't essential).

2. Add the water and the vegetable oil to the transparent container. Leave them to separate out, until the oil is floating on top of the water.

3. Add a completely frozen ice cube to the oil. It should float near the top.

4. Watch as the ice begins to melt, and the denser water drips down through the oil to join the rest of the water below.

5. Eventually, so much of the ice will have melted that the weight of the water will drag the whole cube down, though it might rise again if water drips off the cube.

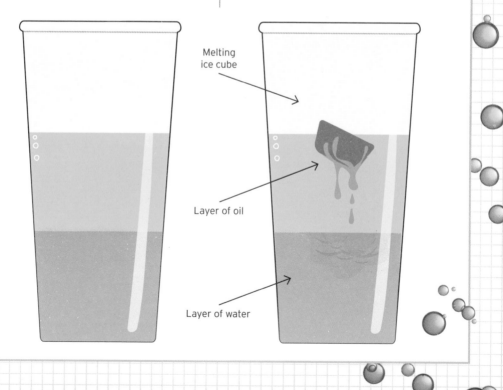

Melting ice cube

Layer of oil

Layer of water

DISCOVER: IN A SPIN OVER HOT WATER

You are likely to find lots of different heating appliances in your kitchen, and none is more ingenious than the microwave oven. It uses some pretty clever chemistry to heat food in a matter of minutes. This section takes a closer look at marvelous microwaves.

First, we need to understand that microwaves are invisible waves that occupy part of what is known as the electromagnetic (EM) spectrum. The EM spectrum is a range of waves of different energies that runs from low-energy radio waves, through the different colors of visible light, to high-energy, dangerous gamma rays.

You might think that microwaves are high energy because of their cooking power, but in actual fact they have only a little more energy than the radio waves that surround us every day. So, if these are low-energy waves, how can they have such a dramatic effect? To understand this, let's look at water molecules again.

POLAR MOLECULES

As a reminder, water is one oxygen atom bonded to two hydrogen atoms. When atoms are bonded together, there is often a tug-of-war over the electrons, and this is particularly true in water. The oxygen atom is greedy for the negatively charged electrons, and pulls them away from the hydrogen atoms. Oxygen isn't strong

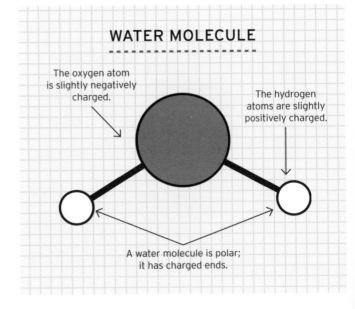

WATER MOLECULE

The oxygen atom is slightly negatively charged.

The hydrogen atoms are slightly positively charged.

A water molecule is polar; it has charged ends.

enough to pull the electrons away entirely from the hydrogen (a process known as ionization), but it is strong enough to keep it closer to itself. This means that the water molecule becomes polar, with a slight negative charge at one side and a slight positive charge at the other.

Something interesting happens when polar molecules are exposed to microwaves. The molecules follow the constantly changing direction of the wave, flipping over and over as they try to align themselves.

The result is water molecules that spin two billion times a second. Since thermal energy is just a measure of how rapidly the particles are moving in a material, this means that microwaves make water very hot, very quickly.

MICROWAVE VS. OVEN

What microwaves make up for in speed, they lose when it comes to quality. Food in a regular oven is cooked at over 100°C (212°F). This temperature is high enough to allow the Maillard reaction to take place, which is a reaction of sugars and amino acids that causes the browning of food and enhances the taste. Because microwaves use water-based heating, it is difficult to get food over 100°C since the water will evaporate, which means the Maillard reaction isn't possible.

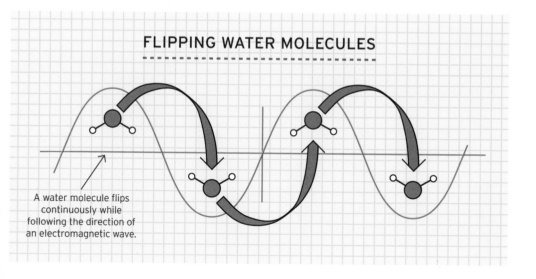

FLIPPING WATER MOLECULES

A water molecule flips continuously while following the direction of an electromagnetic wave.

EXPERIMENT: UNDER PRESSURE

For soft drinks to be filled with fizz, they have to be kept under pressure, but nature doesn't really like differences in pressure. Areas of different pressure will equalize wherever they can: balloons deflate, bike tires need to be pumped up, and ears might pop on a plane.

As you open a bottle of cola and hear that distinctive hiss, the pressure in the bottle has already been balanced with the air pressure in the room. An unopened bottle of cola is very difficult to crush because of the pressure that is sealed inside; it becomes a lot easier as soon as you open it.

In this experiment you'll crush a plastic soft drink bottle, using only the air pressure of the room.

YOU WILL NEED:

- Empty plastic soft drink bottle
- Funnel
- Large plastic bowl
- Ice cubes
- Half a cup of hot (though not boiling) water
- Pitcher of ice water

WHAT TO DO:

1. Stand the empty plastic bottle upright in the bowl and put the funnel in the top. Pour half a cup of hot water through the funnel and into the bottle, forming a layer of water around 2.5 cm (1 in.) deep.

2. Leave the bottle for a couple of minutes, so that the hot water warms the air in the bottle and the pressure inside and outside the bottle equalizes, and then screw the cap on the bottle.

3. Put a layer of ice cubes in the bowl.

4. Lay the bottle down in the bowl and cover it with more ice cubes. Then, pour a pitcher of ice water over it.

WHAT HAPPENS?

The bottle should quickly collapse. The ice and cold water cool the warm air inside the bottle, causing the particles of gas in the air to bounce around less vigorously, reducing the pressure they exert on the walls of the bottle. The pressure of the air outside the bottle is now stronger than the pressure inside, and this pressure difference is enough to crush the bottle.

EXPERIMENT: THE CARTESIAN DIVER

Cola comes in pressurized containers, which are great for keeping soft drinks fizzy, but they do need to be treated carefully. Scientists have been conducting experiments to understand pressure for hundreds of years, and this has given rise to one of the oldest scientific novelties— the Cartesian diver.

The Cartesian diver experiment was first described almost 400 years ago, but it remains one of the simplest and most visual ways of exploring the physical effects that changes in pressure can have. In this experiment, you will build a small diving device that will sink as a response to increasing pressure. The first step in this experiment will be to make the diver.

YOU WILL NEED:
- Pen cap
- Modeling clay or sticky tack
- Empty 2-liter soft drink bottle
- Water

WHAT TO DO:
Making your diver

1. Using the clay or tack, seal the top end of the cap (not the end the pen goes into).

2. Add some clay or tack to the bottom end of the lid—a pea-sized amount should be enough—taking care not to seal the bottom end. This second piece of clay makes the diver heavier.

Testing your diver:

1. Fill the empty bottle to the top with water. Place the cap in the water with the open end pointing down. A bubble of air that is trapped in the cap should keep it afloat.

2. If it sinks, remove some of the clay; if it floats easily, add more clay. You want to have the pen cap just barely floating.

3. Making sure that the bottle is still as full of water as possible, screw the bottle cap back on.

4. Gently squeeze the sides of the bottle and watch what happens.

WHAT HAPPENS?

The diver should, if you squeeze enough, sink down to the bottom of the bottle. If you stop squeezing, it should make its way back to the top.

There are two important factors at work here: the Boyle-Mariotte law and Archimedes' principle. The Boyle-Mariotte law tells us that as the pressure on a gas is increased, its volume decreases. This means that as you squeeze the bottle and increase the pressure, the volume of the gas in the air bubble trapped inside the pen cap decreases. The trapped bubble gets smaller.

Archimedes' principle states that the buoyant force (the force allowing the cap to float) is equal to the weight of the liquid which has been displaced by the object.

You can understand what is happening by combining these effects. The increase in pressure leads to a decrease in bubble size, which means that less water is displaced, and the cap experiences a smaller buoyant force, so it sinks.

Reducing the pressure allows the air bubble to expand, increasing the buoyant force and sending the cap back to the top.

LEARN ABOUT: TESTING TEMPERATURES

How hot is your knowledge of temperature?

1. When ice is added to a drink the liquid becomes colder. Why?
a) Thermal energy is absorbed by the ice cube.
b) The liquid absorbs some of the coldness from the ice.
c) The ice melts, which dilutes the warm liquid.

2. If you put ice cubes in a glass and fill it to the brim with cola, what will happen when the ice melts?
a) Nothing—the glass remains full.
b) The glass becomes too full and cola spills out.
c) The glass becomes less full.

3. Substances can be transformed from solid to liquid, and from liquid to gas, by being heated. Therefore, gases are always hotter than liquids. True or false?

4. How much less dense is ice than liquid water?
a) 3.5 percent
b) 8.3 percent
c) 12.8 percent
d) 15.2 percent

LEARN ABOUT: TURNING UP THE HEAT

Try these questions on heating, cooling, and the states of matter.

POP QUIZ: HEATING AND COOLING

1. What doesn't happen when water evaporates?
a) It increases in volume.
b) It has a lower density.
c) The particles move with more freedom.
d) Chemical bonds between atoms are broken.

2. Ketchup is a non-Newtonian fluid because:
a) It becomes less dense when it freezes.
b) It doesn't crystallize when frozen.
c) Its density changes as a result of applied stress.
d) It cannot be compressed.

3. Rank these from most to least dense under normal conditions. *(Hint: Which would float on the others?)*
a) Air
b) Water
c) Ice
d) Vegetable oil

4. When a gas is compressed, do the following increase or decrease?
a) Volume
b) Pressure
c) Density
d) Space between the particles

5. What is Archimedes' principle?
a) The buoyant force is equal to the weight of the liquid displaced.
b) An object of high mass can float if it has a high surface area.
c) Under water, a gas will be compressed more as depth increases.
d) Objects appear larger when submerged.

6. Which of these statements describes a polar molecule?
a) It contains atoms that have different numbers of protons.
b) It has a charged atom.
c) The electron distribution is not even, resulting in localized charges.
d) It is magnetic.

7. What happens to a water molecule exposed to microwaves?
a) It is ionized and separates into charged parts.
b) Its bonds stretch and bend.
c) Nothing; the particles do not react.
d) It spins around.

8. Glass does not get hot in microwaves. Why is this?
a) Microwaves are designed to interact with polar water molecules; without polar molecules in the glass, there is very little for the microwaves to interact with.
b) Glass particles are not free to move and therefore do not heat effectively.
c) Glass is not a good heat conductor.
d) Glass is transparent, so microwaves pass straight through the material.

9. Why is the Maillard reaction not possible with microwave cooking?
a) Food is heated from the inside out, meaning the surface is the last to cook.
b) Microwave ovens use water-based heating and the temperature doesn't get high enough.
c) Microwaves cut out when burning is detected.
d) The Maillard reaction happens slowly and the duration of microwave heating is too short.

CHAPTER 2
SOLUTION

DISCOVER...

LEARN...

EXPERIMENT...

DISCOVER: ARE YOU CONCENTRATING?

People around the world enjoy cola's sweetness and flavors, complemented by its thousands of bubbles. The balance of the ingredients is vital: a little more or less of any essential ingredient, and the drink might be ruined. The key lies not just in the ingredients, but in their concentrations.

A solution is a substance (the solute) dissolved in a liquid (the solvent). Although it varies from country to country, 330 ml (11 fl. oz.) of cola is typically a solution containing about 35 g (7 tsp.) of sugar. (Sugar is the solute, water is the solvent, and cola is the solution.)

Calculating concentrations is one of the most fundamental and important concepts in chemistry, and there are lots of ways to do it. Concentration tells us how much of something there is compared to the solution it is in. This is important for chemists to know because it indicates how a solution will react. It's important for beverage manufacturers to know because it will determine how a drink tastes!

CONTAINS 35 GRAMS OF SUGAR

In its simplest definition, the concentration is equal to the amount of the solute (e.g., sugar) divided by the volume of the solution (e.g., the bottle of cola). In reality, we can think about concentration in different ways, depending on what will be most useful.

You could work out the concentration by taking the number of grams of solute and dividing it by the number of liters of the solution. This would give you a concentration in grams per liter (g/L). Or you might want to know how many moles are in a solution (discover moles on pages 42–43).

WATER AND JUICE

Imagine you have two barrels, one containing water and one containing fruit juice. You take a cup of the water and add it to the barrel of juice. You then take a cup of the water–juice solution and add it to the barrel of water. Both barrels start and finish with the same volume, but which is the purer solution? (i.e., which has the greater proportion of just one substance?)

The water? Think again.
The juice? Nope.

Both barrels have the same purity!

It may help to think about this with marbles instead: 100 red and 100 white. You take a cup of the red marbles and add them to the white marbles. You then take a cupful of the red–white mix and add it to the red. Let's say the cup holds 20 red marbles which you added to the white ones. The white barrel now has 120 marbles, so you need to take 20 back out at random to balance the barrels. When you fill the same cup from the white barrel, let's say you randomly collect 15 white and 5 red. Now you have one barrel with 85 white and 15 red and another with 85 red and 15 white. The barrels will always have swapped the same number of marbles with the other barrel, and so the purity of the mixes will always be the same.

LIQUID LANGUAGE

Solutes don't have to be solids. If you are mixing two liquids, the one with the smaller volume is the solute, while the larger is the solvent.

MIXING THINGS UP

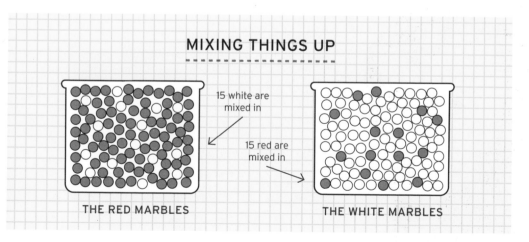

15 white are mixed in

15 red are mixed in

THE RED MARBLES

THE WHITE MARBLES

DISCOVER: MASSIVE MOLES

Moles are everywhere in chemistry. Wherever you go, and whichever field you work in, you are likely to be surrounded by them. But chemistry moles aren't furry little mammals—they're a way of quantifying different substances.

Chemists have to deal with complex reactions, and they need to know not just the mass or volume of their chemicals, but how many particles they contain. This is where moles come in.

One mole of material contains 6.022×10^{23} particles—what is known as Avogadro's constant. For example, one mole of gold, which has a mass of 196.97 g (about 7 oz.), contains 6.022×10^{23} atoms. This gives chemists a really useful way to compare substances that might be very different.

The number 6.022×10^{23} is huge, so why is it helpful? Reactions take place on a small scale, and you need to know exactly how many atoms or molecules substances have in order to understand what they will react with and how. Moles help us think about how many particles are in the material on a scale we can see and measure.

STRIKING THE RIGHT BALANCE

Let's say you want to produce a reaction between hypothetical molecules A and B. Each A pairs with one B and forms a new molecule. If you wanted to produce this hypothetical reaction between helium and gold, with each gold atom reacting with one helium atom, and took the same mass of the materials—10 g of gold and 10 g of helium—you would encounter a problem. You'd have many more atoms of helium than gold because the mass of a helium atom is less than the mass of a gold atom. To find the correct amount to mix, you need to convert the masses into moles!

Italian scientist
Amedeo Avogadro
(1776-1856)

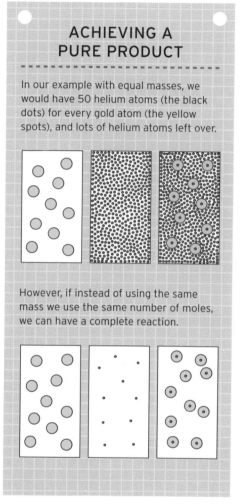

ACHIEVING A PURE PRODUCT

In our example with equal masses, we would have 50 helium atoms (the black dots) for every gold atom (the yellow spots), and lots of helium atoms left over.

However, if instead of using the same mass we use the same number of moles, we can have a complete reaction.

HOW MANY MOLES?

To calculate the number of moles in a substance, we must know the relative mass of the particles. Every atom has its own relative atomic mass, which is the average mass of the atom compared with one twelfth of a carbon atom. When we add up the relative atomic masses for all the elements in one particle, we get the relative formula mass.

Let's take water as an example. The formula of water is H_2O, so we have two hydrogen atoms and one oxygen atom. The relative atomic mass of hydrogen is 1, and the relative atomic mass of oxygen is 16, so the relative formula mass of water is $1 + 1 + 16 = 18$. One mole is equal to the relative formula mass in grams, meaning that one mole of water has a mass of 18 g (½ oz.), and 18 g of water contains 6.022×10^{23} water molecules.

If you want to work out the number of moles you have, there is a simple formula to use:

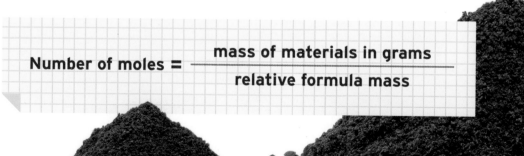

$$\text{Number of moles} = \frac{\text{mass of materials in grams}}{\text{relative formula mass}}$$

DISCOVER: ACIDS

When acids make an appearance in horror movies, they tend to be shown as terrifying vats of bubbling liquid that can dissolve anything in seconds. In reality, acids tend to be a little less dramatic, but no less interesting.

If you look closely in your kitchen, you're bound to find lots of acids. The vinegar in the cupboard is a dilute solution of ethanoic acid. Orange juice gets its distinctive flavor in part from citric acid, and it contains ascorbic acid, also known as vitamin C. In cola, you're likely to find phosphoric acid, citric acid, and carbonic acid—none of which can dissolve things in seconds, but they could still be a little horrifying for your teeth.

WHAT IS AN ACID?

If you ask a chemist what an acid is, you might get a few different answers because the term covers a range of chemical properties. The most common and simplest definition of an acid is a molecule or ion capable of donating a proton (a hydrogen atom which has lost its electron, H^+).

An example of acid behavior would be hydrogen chloride (HCl) in water, which makes hydrochloric acid. When you add HCl to water, it splits into chloride ions (Cl^-) and H^+ and these are very reactive. The H^+ will react with water and form a hydronium ion, H_3O^+. The hydrogen chloride has given the water a proton, so it has behaved as an acid.

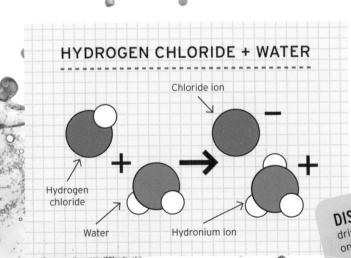

HYDROGEN CHLORIDE + WATER

Chloride ion

Hydrogen chloride

+

Water

→

+

Hydronium ion

+

DISCOVER the effect acidic drinks can have on your teeth on pages 56–57.

FEEL THE BURN

Some acids can be very harmful to humans because the H^+ and H_3O^+ are very reactive. There are two main things which tell us how dangerous an acid will be: its strength and its concentration.

Whether an acid is classed as strong or weak depends on how easily split apart the acid is. Hydrochloric acid will almost entirely split when it is added to water, meaning there won't be any molecules of HCl in the water, but instead Cl^- and H^+ or H_3O^+ ions. Hydrochloric acid is a strong acid. With weak acids, at any one time only some of the molecules will have split into ions.

The second important thing you need to know about an acid is how concentrated it is. A high concentration of hydrochloric acid would burn you, but the effects of diluting the same amount in a swimming pool would go unnoticed. Most of the acids you find around the house are weak and dilute, which means they aren't very dangerous. For example, white vinegar, found in many homes, is a dilute solution of a weak acid (ethanoic acid), making it safe for cleaning and cooking.

And even though cola is an acidic mix of three different acids, it's no match for the strong, concentrated hydrochloric acid in your stomach!

Cola contains phosphoric acid, citric acid, and carbonic acid.

DISCOVER: DROP THE BASE

A big dinner of pickles and fries, with lots of malt vinegar, washed down with a bottle of cola and a glass of orange juice? These acidic ingredients will start to brew up a sore stomach, and may have you reaching for an antacid!

When our stomachs have too much acid, the acid can creep up the esophagus, causing heartburn. An antacid will react with the acid and neutralize it (often this reaction produces gas, making you burp).

NEUTRALIZING ACIDS

To neutralize acids, chemists use a base. If an acid is something that can donate a proton (H^+), a base is the opposite: it can accept a proton. To go back to the hydrochloric example from page 44, an HCl molecule splits into a proton (H^+) and a chloride ion (Cl^-). In order to neutralize the acid, the reactive ion needs to be removed, and this is what the base does.

Sodium hydroxide (NaOH) dissolved in water makes a good base. When sodium hydroxide is added to water, it splits into sodium ions (Na^+) and hydroxide ions (OH^-). The hydroxide ions are very reactive and easily react with the protons found in acid. By dissolving NaOH in water, a solution is created that will accept protons and therefore neutralize acid.

When a base is dissolved in water, the solution is called alkaline. When hydrogen chloride is dissolved in water, the solution is acidic, whereas the sodium hydroxide solution is alkaline.

A BASE IN YOUR BAKING

Baking powder is used to make cakes rise. It does this through an acid and base neutralization reaction. The reaction produces bubbles of carbon dioxide, creating the sponge structure.

SALTY SCIENCE

If the acidic solution and the alkaline solution are mixed, they will react with each other. The H^+ from the acid reacts with the OH^- to form H_2O (water). And the Cl^- and Na^+ react to form $NaCl$, or table salt.

If the balance is right and there is the same number of moles of acid and base (see pages 42–43), all of the hydrochloric acid is neutralized, leaving just salt water.

But what would happen if the math was wrong and too much sodium hydroxide was added? It would react all of the H^+, so the solution would no longer be an acid, and lots of OH^- would remain in the solution. This means the solution would have changed from being acidic to alkaline.

Even though sodium hydroxide could technically neutralize excess acid in your stomach, it would be very harmful. Alkaline solutions are just as dangerous as acidic ones—a solution of sodium hydroxide can burn skin. To remedy upset stomachs, something much gentler is needed from the pharmacy, such as aluminum hydroxide or magnesium carbonate.

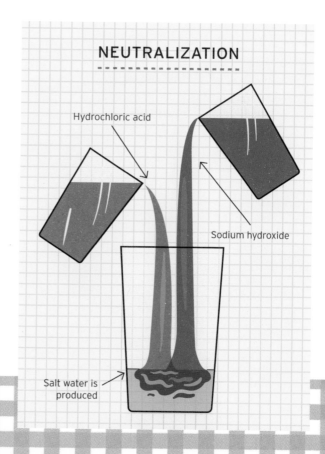

NEUTRALIZATION

Hydrochloric acid

Sodium hydroxide

Salt water is produced

DISCOVER: ACID REACTIONS

You might have heard the expression, "opposites attract." In this case, opposites react. Let's take a closer look at how to measure the acidity of a liquid and what happens when acids react.

You can find lots of acids and bases at home: cola is acidic and bleach is basic; vinegar is acidic and egg whites are basic. How acidic is cola, and how basic is bleach? We need to be able to compare the acidity of different substances, and this is done using the pH scale.

THE ACID TEST

The pH scale is how scientists measure how acidic or basic a solution is. Given that an acid is a substance that can donate a proton (H^+), we can understand the pH scale as a way of measuring the concentration of H^+ ions in a solution.

The pH scale runs from 0 to 14. Right in the middle of the scale (7) is the neutral point: anything with a pH less than 7 is acidic, and anything with a pH greater than 7 is basic. Pure water is neutral and lies in the middle—it's neither an acid nor a base. Cola has more H^+ ions than water, so it has a lower pH—around 2.5—whereas stomach acid has a pH of 1.5–3.5. It might be surprising that cola is as acidic as your stomach, which is capable of breaking down food. Teeth are particularly sensitive to acidic substances, so cola should be limited to a special treat.

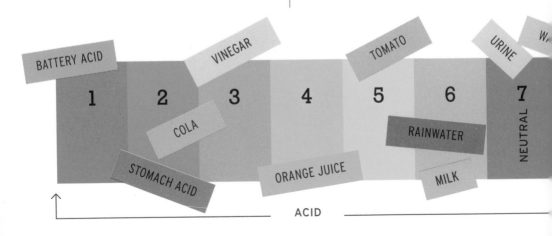

BATTERY ACID

VINEGAR

TOMATO

URINE

W

1   2   3   4   5   6   7

COLA

RAINWATER

NEUTRAL

STOMACH ACID

ORANGE JUICE

MILK

ACID

KITCHEN BASICS

When solutions have a pH higher than 7, they are basic. A lot of the cleaning products you might find at home are basic. Bleach is basic and has a pH around 12.

An even stronger base is drain cleaner, which can have a pH of 13. Both of these are corrosive and should be handled with care. A liquid with a high pH is just as capable of burning skin as an acid.

While some bases can do us harm, others are useful as medicine. Antacid tablets are basic and are used to neutralize excess stomach acid. Though beware: this neutralization reaction can produce lots of gas.

KICKING UP A FIZZ

Here is a neutralization reaction to try at home: put 50 ml (1²/₃ fl. oz.) of vinegar into a glass and then add a tablespoon of baking soda. The mix will begin to fizz and react. The H⁺ is used up and water is produced, with a resulting pH level close to neutral. The reaction also fizzes as it gives off carbon dioxide—the gas that gives cola its fizz.

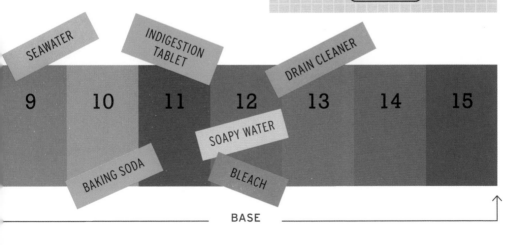

SEAWATER INDIGESTION TABLET DRAIN CLEANER

| 9 | 10 | 11 | 12 | 13 | 14 | 15 |

SOAPY WATER BAKING SODA BLEACH

BASE

EXPERIMENT: ACID DETECTIVE

Chemists want to understand the world around them. Conducting experiments and recording observations helps them to do so. In this experiment, you will make a pH indicator using red cabbage and test the pH of several household chemicals.

Whether something is classed as an acid, a neutral, or a base depends on the concentration of H^+ ions. One of the most common ways of identifying which category a liquid falls into is by adding an indicator.

A pH indicator is something that will produce a visible change, reflecting the pH of the solution it is added to. This allows you to tell if something is an acid, neutral, or base simply by looking at it. A good indicator will even tell you what the pH of the liquid is.

ADULT SUPERVISION REQUIRED

YOU WILL NEED:

- Red cabbage leaves
- Blender or food processor
- Sieve
- 500 ml (1 pt.) of water
- Measuring pitcher
- Several small glasses

SUGGESTED TESTABLE SUBSTANCES:

- Vinegar
- Lemonade
- Lemon juice
- Baking soda
- Liquid soap
- Antacid tablets
- Laundry detergent

SAFETY FIRST You may have stronger acids and bases at home, but remember that these can be harmful and should be treated with care. Avoid direct contact with strong acids or bases.

WHAT TO DO:
Part one: Prepare the indicator

1. Roughly chop the cabbage leaves and add them, along with the water, to the blender. (Ask an adult for help.)

2. Blend the cabbage and water until the leaves are broken down and the mixture looks smooth.

3. Pour the mixture through the sieve into the measuring pitcher. The water will be a purple-blue color.

4. Pour some of the liquid into the small glasses, with about 50 ml (1½ fl. oz.) in each. Use as many glasses as you have substances you want to test.

Part two: Test the pH

1. The water you have in the glasses is neutral. This is the indicator. If you add an acid or a base, the liquid should change color.

2. Carefully add several drops of the liquid you want to test to the small sample of cabbage water, and mix.

3. Watch what happens to the color of the liquid. It can help to see any color changes by placing the glass on a piece of white paper. (If you want to test a solid such as an antacid tablet, crush it and then dissolve it in the water.)

4. If you have added an acid to the indicator, the water should turn red; if you have added a base, it should turn blue. A strong base might even turn the liquid green.

5. Line the samples up from acidic, through neutral, to basic.

THE TESTING KIT

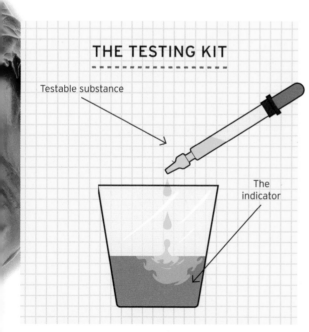

Testable substance

The indicator

WHAT HAPPENS?

This experiment works because cabbage contains chemicals that react in different ways, depending on the concentration of H^+ ions in the testable ingredients. This allows you to compare the pH of household goods.

DISCOVER: SUPER SOAP

It might not seem obvious, but some of the most mundane household objects are superheroes in disguise. Soap is one such hero: we use it every day without appreciating quite how special it is. The chemistry behind how it works is pretty remarkable too.

Soap is probably the oldest and simplest of the cleaning products you have in your house. It is chemically defined as the salt of a compound known as a fatty acid. Fatty acids are the building blocks of fats and are composed largely of a chain of carbon atoms bonded with hydrogen atoms. This definition helps to understand how soap cleans.

CLEANING PROPERTIES

Soap has a long carbon chain, made of carbon and hydrogen atoms, which does not easily mix with water. This chain is described as hydrophobic, or "water fearing."

Soap is also a salt, which is a substance made of ions (charged atoms or molecules). Soap normally contains sodium, with the sodium losing its outer electron to the fatty acid. This leaves the sodium positively charged and the fatty acid negatively charged. When we mix soap with water, the sodium is free to move around. The positively charged sodium moves away from the fatty acid chain, leaving a negatively charged area behind. This negatively charged site is attractive to the water and is described as hydrophilic, or "water loving."

Soap, then, has a long chain that doesn't like water, but which has a negatively charged end (called the head), which loves water.

INGENIOUS SOAP

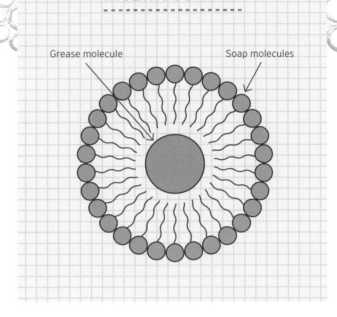

Grease molecule

Soap molecules

The next step in understanding how soap cleans is to work out what makes our hands or surfaces dirty in the first place. In most cases, it will be grease, oil, or dirt, which are made mainly from carbon and hydrogen atoms. This carbon-based dirt will also be hydrophobic, meaning that it can't easily be washed away because it repels the water from its surface.

SOAP SAVES THE DAY

There is a saying in chemistry: "Like dissolves like." What that means in this case is that water-repelling oils can dissolve water-repelling stains, which is useful. But while we might be able to remove household dirt with oil, it would also leave our hands, clothes, and worktops covered in oil. What we need is something that can remove the carbon-based dirt and then be washed off with water. In other words, what we need is soap!

When soap molecules meet carbon-based dirt, oil, or grease, they surround it. The soap molecules line up around the dirt with their water-repelling tails facing the dirt and their water-loving heads pointing out toward the water. It is as if the dirt becomes surrounded by a water-loving bubble. This means the oil and dirt, which wouldn't normally dissolve in the water, can now be washed away.

LEARN ABOUT: MOLE HUNTING

Moles are really important to getting experiments right. If you know how many moles of something you have, you can work out how many particles are in your sample. Now it's time to test what you've learned!

Before you start, here are some useful reminders of the different terms:

Relative atomic mass (A_r): The average mass of an atom compared with one twelfth of a carbon atom. Since atoms have such a tiny mass, they are compared with carbon to make the number easier to work with.

Relative formula mass (M_r): The sum of all the relative atomic masses in the molecule. For example, methane has the formula CH_4; to calculate M_r, we add up the A_r for each of the atoms. Carbon has an A_r of 12 and hydrogen has an A_r of 1. The M_r of methane is then $12 + 1 + 1 + 1 + 1 = 16$.

Avogadro's constant: This is the number of constituent particles in one mole. It is a constant with a value of 6.022×10^{23}.

POP QUIZ: MOLES

1. Sucrose is a sugar commonly found in soft drinks. It has the formula $C_{12}H_{22}O_{11}$. The relative atomic mass of carbon (C) is 12, for oxygen (O) it's 16, and for hydrogen (H), 1. What is the relative formula mass (M_r) of sucrose?

2. If a can of cola has 11 g of sucrose, how many moles of sucrose does it contain? *Hint: Page 43 has an equation which is useful.*

3. You measure out 100 g of sucrose and 100 g of water. Which of the samples contains the most molecules?

4. An eccentric septillionaire decides to buy one mole of her favorite candy, which comes in solid 1 cm³ cubes (which here represents the atoms). She lays them side by side all across the United States, which has an area of 9.83×10^{16} cm². Could she cover all of the US? *Hint: The base of the cube will be 1 cm². Which has a larger area, one mole of the candy or the United States?*

LEARN ABOUT: A BURNING QUESTION

Now that you've explored the world of acids and bases, try this test of your pH mettle.

POP QUIZ: ACIDS AND BASES

1. An acid is usually defined as:
a) a corrosive liquid
b) a substance that can donate a proton
c) a substance that gives off heat when reacting
d) a liquid capable of ionizing water

2. The pH of milk is approximately:
a) 3.2
b) 6.6
c) 7.0
d) 8.4

3. pH is a measure of:
a) how easily H+ ions dissociate (split) from a molecule
b) the charge of the nonwater component
c) the threat to human tissue
d) the concentration of H+ ions

4. Which of the following statements is true?
a) Rainwater is neutral and seawater is basic.
b) Rainwater is basic and seawater is acidic.
c) Rainwater is acidic but seawater is basic.
d) Rainwater is acidic and seawater is neutral.

5. A strong acid is one which:
a) has a high concentration
b) is fully ionized
c) is only partially ionized
d) has a high reaction rate

6. Which of these has the highest pH?
a) Vinegar
b) Cola
c) Soapy water
d) Pure water

7. You add five drops of solution A to beaker B and the pH in the beaker increases. The solution is:
a) acidic
b) alkaline (basic)
c) neutral
d) not possible to determine

8. If a liquid has a pH of 10, it is:
a) basic
b) acidic
c) neutral

EXPERIMENT:
TERRIFYING TEETH

Most people know that too many sugary drinks can be bad for their health and for their teeth. But it's easy to forget the damage that can be done to teeth because the process is slow— it can take many years for problems to occur. This experiment, speeds up the process to show what can happen to teeth.

There are two things that cause sugary drinks to damage teeth: sugar and acid. The pH of your mouth is normally around neutral, which is a healthy environment for teeth. If the pH is lowered and the mouth becomes acidic, the teeth start to lose calcium, which softens the enamel and leads to cavities.

The sugar is also linked to acid levels, but less directly. When you eat or drink something with sugar, you not only feed yourself, but you feed the millions of bacteria that live in your mouth. Sugar is a great food for bacteria: it can be turned very quickly into energy. During this process the bacteria will produce lactic acid. The lactic acid produced by the bacteria in your mouth then leaches calcium from your teeth, just like the phosphoric and citric acids in cola.

Let's look at this process in action.

YOU WILL NEED:
- 500-ml bottle of cola with a screw-on cap
- At least 1 baby tooth—if you don't have a baby tooth to use, you could use a chicken bone
- Sewing needle
- Sieve
- Measuring pitcher
- Camera
- Pen and paper

WHAT TO DO:

1. Place the tooth on a flat surface with a uniform color—using a piece of white paper or colored cardboard works well—and photograph the tooth. Take the photograph somewhere the light is always the same.

2. Using the sewing needle, examine the tooth. Is it possible to scratch the tooth? If you try to gently press the needle into the tooth, what happens? Is the root of the tooth soft or hard? Write your answers down.

3. Put the tooth in the bottle of cola and replace the cap. Store the bottle somewhere dark, on a shelf or in a cupboard. Please bear in mind that this experiment will likely destroy the tooth, so don't use any cherished childhood keepsakes!

4. After three days, pour the bottle of cola out, through the sieve into the measuring pitcher. Examine the tooth with the needle, and take a photograph of it. Put the cola and the tooth back into the bottle, and reseal.

5. One week after the tooth was originally added, repeat step 4. Can you see any differences? You might want to replace the tooth and run the experiment for longer—for a month, or more.

DISCOVER MORE
about sugar on pages
92–93.

WHAT HAPPENS?

You will probably find that the tooth has become discolored. The roots, being the thinnest part and lacking hard enamel, will become softer. After several weeks, you will be able to easily scratch the tooth.

If you have other spare baby teeth, try other liquids. Try sugar-free cola; is there a noticeable difference? You might also want to try water or milk.

DISCOVER: CHEMICAL MATH

When chemicals react or change, we can often see the results, but we can't see the individual chemicals reacting. Chemists need to understand and communicate what is happening at the level of the individual particles. This is where chemical equations come in.

Chemical equations represent what happens in a reaction. The reactants are written on one side (the starting ingredients) and the products (the chemicals formed from the reaction) are written on the other side. Just like in math, the equations need to be balanced, which means having the same number and type of atoms on each side.

Here are a few steps to help you understand the world of balanced chemical equations.

1. Pointing the way

One of the biggest differences between a chemical equation and a mathematical equation is that there is no equal sign in a chemical equation. Instead, chemists use arrows. This helps show how the reaction proceeds—it tells us what is reacting and what is being formed.

2. Two sides to every story

In a chemical equation, the reactants and products are on opposite sides, with an arrow pointing from the reactants to the products:

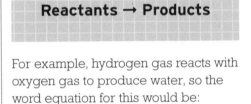

Reactants → Products

For example, hydrogen gas reacts with oxygen gas to produce water, so the word equation for this would be:

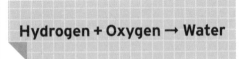

Hydrogen + Oxygen → Water

3. Space-saving symbols

Chemists write the reactants and the products in symbol form, using the symbols from the periodic table (see pages 68–69). Hydrogen has the symbol H and each molecule of hydrogen contains two hydrogen atoms, so we write H_2. The symbol for oxygen is O, and each oxygen molecule also has two atoms, so we write O_2.

Now we have the equation:

$$H_2 + O_2 \rightarrow H_2O$$

4. Finding the balance

The equation above has a problem: the two sides aren't balanced. On the left there are two hydrogen atoms and two oxygen atoms, whereas on the right there are two hydrogen atoms and one oxygen atom. It's important to have the same numbers on each side; we can do this by adding more of what we need. For this equation to balance, two hydrogen molecules need to react with a single oxygen molecule.

Our water-producing equation has become:

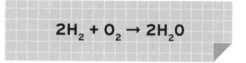

$$2H_2 + O_2 \rightarrow 2H_2O$$

This means we have four hydrogen atoms and two oxygen atoms on the left and right sides. The equation is balanced! Two molecules of hydrogen will react with one molecule of oxygen to produce two molecules of water.

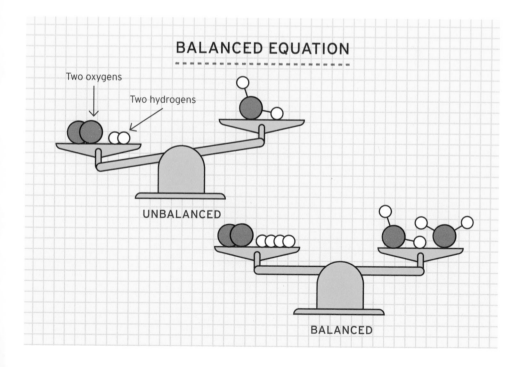

BALANCED EQUATION

Two oxygens

Two hydrogens

UNBALANCED

BALANCED

LEARN ABOUT: THE CORRECT BALANCE

The process of balancing an equation makes clear how much of one reactant is needed compared with another in order to achieve the desired product. Pages 58–59 explained how to form a balanced equation, and now it is time to put those skills to the test.

POP QUIZ: BALANCING EQUATIONS

1. The balanced equation for hydrogen reacting with oxygen to produce water is $2H_2 + O_2 \rightarrow 2H_2O$. What would happen if you reacted one mole of hydrogen with one mole of oxygen?

2. All balanced equations need to have the same number of atoms on both sides. What does this mean?
a) The total mass of the reactants equals the total mass of the products.
b) The reaction is reversible.
c) The products are just as reactive as the reactants.
d) The total volume of the reactants equals the total volume of the products.

3. Which common molecule is missing from this equation, which shows the burning of ethane? Fill in the blank:

$2C_2H_6 + 7O_2 \rightarrow 4CO_2 + 6$ ___

4. Chemists write the state of the reactants and products after their symbols: solids (s), liquids (l), gases (g), or aqueous, dissolved in water (aq). Add the correct states to the following equation, which shows metallic gold reacting with a cloud of chlorine to produce yellow crystals of gold chloride.

$2Au(\ ) + 3Cl_2(\ ) \rightarrow AuCl_3(\ )$

5. Balance the following equations. Each one is trickier than the last!

a) $TiCl_4 + H_2O \rightarrow TiO_2 + HCl$
b) $Fe + O_2 \rightarrow Fe_2O_3$
c) $C_{12}H_{22}O_{11} + O_2 \rightarrow CO_2 + H_2O$

6. Write a word equation for the following reaction (you might recognize this reaction from pages 46-47):

$HCl + NaOH \rightarrow NaCl + H_2O$

7. What information does a balanced equation not usually give us?
a) The products of a reaction
b) Which state the reactants are in
c) The molar ratio of the reaction
d) The rate of reaction

8. There is always a higher number of reactants than there are products. True or false?

LEARN ABOUT: THE LAB

There are similarities between the equipment and safety warnings found in a laboratory and those you might find at home. Match each piece of lab equipment and warning symbol (1–9) to its correct description (A–H).

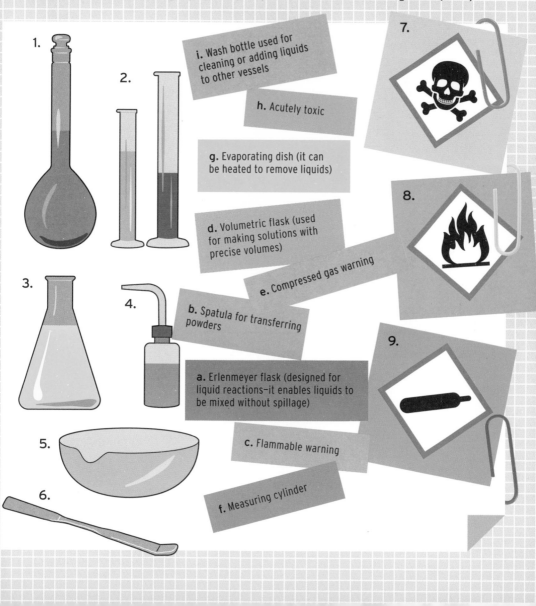

1.

2.

3.

4.

5.

6.

7.

8.

9.

i. Wash bottle used for cleaning or adding liquids to other vessels

h. Acutely toxic

g. Evaporating dish (it can be heated to remove liquids)

d. Volumetric flask (used for making solutions with precise volumes)

e. Compressed gas warning

b. Spatula for transferring powders

a. Erlenmeyer flask (designed for liquid reactions–it enables liquids to be mixed without spillage)

c. Flammable warning

f. Measuring cylinder

EXPERIMENT: HURRY UP!

Chemists don't just want to understand reactions; they also want to control them. In the kitchen, it is important to know how long something will take to cook, or how long before food goes bad. Behind these everyday occurrences are complex chemical reactions.

To control a reaction, you need to understand the reaction rate—how quickly the products are formed or the reactants used up.

This experiment explores some of the different factors that can influence the rate of a reaction, by testing the time it takes to dissolve a solid in water. Different conditions are compared to see how the reaction speed changes.

YOU WILL NEED:

- Dissolvable fizzy orange vitamin C tablets or sugar cubes
- Hot and cold water
- Stopwatch
- A few glasses

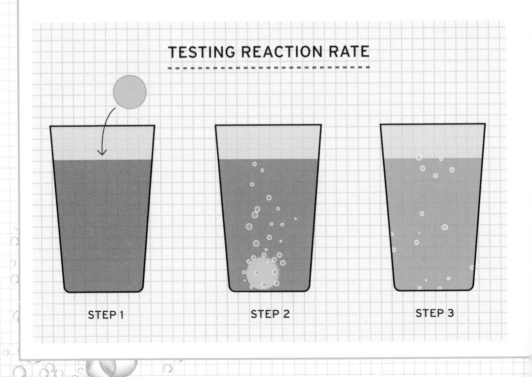

TESTING REACTION RATE

STEP 1   STEP 2   STEP 3

WHAT TO DO:

First conduct a control experiment under what will be termed standard conditions. You will then be able to compare the results of this experiment with ones where you have altered the conditions.

1. Fill a glass with room-temperature water (if the water from the tap is cold, leave it to become room temperature).

2. Drop a dissolvable vitamin C tablet or sugar cube into the water. Start the stopwatch.

3. Allow the experiment to run until the solid is completely dissolved in the solution and record the time it took.

The time it took tells us how long the tablet or cube will take to dissolve under standard conditions.

TEMPERATURE

One of the simplest ways to alter the rate of a reaction is to change the temperature at which it takes place. Repeat the above experiment with cold water and then hot water and compare the time it takes for the solid to completely dissolve.

You should find that the hotter the water is, the faster the reaction will occur. For a reaction to take place, the reactants have to come into contact with each other. Increasing the temperature will allow the particles to move more quickly. The faster they move, the more likely they are to collide with one another and react.

SURFACE AREA

Repeat the control experiment, but this time crush the tablet or cube into a powder before adding it to the water.

The powder should very quickly dissolve. In tablet or cube form, the water only has access to the outside of the cube; if you crush it, you increase the surface area. With a higher surface area, the water can come into contact with much more of the solid from the outset, and the reaction rate is increased.

STIRRING

Repeat the experiment, this time constantly stirring the glass. This should increase the number of collisions between the tablet and the water and again increase the reaction speed.

What happens when you combine the three variables—temperature, surface area, and stirring? You should find the reaction goes even faster.

CHAPTER 3
CHEMICAL MAKEUP

DISCOVER...

LEARN...

EXPERIMENT...

DISCOVER: DIVIDING THE INDIVISIBLE

If you **want to understand the chemistry of anything**—cola, cleaning products, planets—you need to know something about the atoms it is composed of. Atoms are famously called "the building blocks of matter," and everything is made of them!

THE BUILDING BLOCKS OF BUILDING BLOCKS

More than 2,000 years ago, philosophers theorized that the world was made of tiny units. They hypothesized that these tiny units were the smallest units possible and were indivisible, meaning that they could not be broken down into anything smaller. It turns out that their theory was wrong.

After 2,000 years of experimentation, guesswork, and new ideas, scientists were starting to realize that the atom isn't indivisible, but is in fact made of smaller particles. A mere 100 years ago, the modern conception of the atom emerged.

LOOKS CLOUDY

The basic structure of all atoms is the same: a dense nucleus composed of protons and neutrons orbited by electrons. Almost all of the mass of an atom is contained in the nucleus, while the nearly massless electrons whizz around the outside. When chemists draw atoms, they tend to look a little like diagrams of the solar system, with the nucleus taking center stage like the Sun and the electrons forming circular orbits at some distance from the high-mass center.

The solar system analogy helps us visualize things, but the truth is a lot stranger.

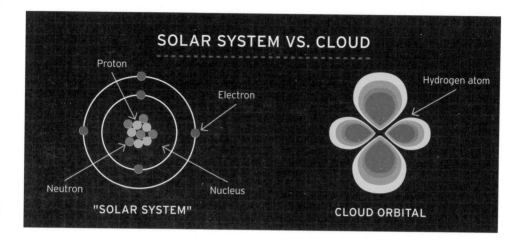

SOLAR SYSTEM VS. CLOUD

Proton

Electron

Hydrogen atom

Neutron

Nucleus

"SOLAR SYSTEM"

CLOUD ORBITAL

That analogy is flawed. Electrons do orbit the nucleus, but not like a planet. The very small, very light electrons travel so fast that knowing exactly where they are at any point is impossible. Instead, the electrons can be thought of as a cloud around the nucleus. Chemists and physicists call these clouds orbitals. The simplest orbital is a sphere around the nucleus. As we add more and more electrons, protons, and neutrons into the atom, the orbitals become more complex, forming lobed structures pointing in different directions.

THE SAME BUT DIFFERENT

All atoms follow the same basic blueprint, but as protons, electrons, and neutrons are added, the properties of the atom change. The different types of atoms, known as elements, are determined by the number of protons in their nucleus. For example, atoms with 17 protons are chlorine, and atoms with 18 protons are argon. The number of protons in a nucleus is known as the atomic number, and chemists have observed 118 different atomic numbers, meaning 118 different elements have been discovered or created in the laboratory.

THE SUBATOMIC PARTICLES

Atoms are made of smaller particles: electrons, protons, and neutrons.

ELECTRON: a negatively charged particle

PROTON: a positively charged particle with a mass around 2,000 times that of an electron

NEUTRON: an uncharged particle of similar mass to a proton

DISCOVER: THE PERIODIC TABLE

The products in the soft drinks aisle in the supermarket are arranged so that shoppers can easily find what they want. The periodic table does a similar thing for chemists—it sets out every element in a very clever way, so it's easy to see how they behave and what properties they have.

An element is defined by how many protons are in the nucleus of the atom. This is known as the atom's atomic number. Each element has a different atomic number—from element number one, hydrogen, through to element number 118, oganesson. The table opposite might look complete, but as scientific abilities advance, it will be possible to create heavier and heavier elements, and chemists will probably have to add a new row to the table in a few years' time.

TOWERS OF SIMILARITY

When you read text in English, you move from left to right. When you read the periodic table, you can read upward, downward, across, or even diagonally. The columns of the periodic table are called groups, and elements in the same group behave in similar ways. For example, as you look at the left-hand side—the group that includes lithium, sodium, and potassium—all of these react violently with water. Or, on the right-hand side—the group with neon, argon, and krypton—all of these are inert gases, meaning that they are unreactive. Elements in the same group behave in similar ways because they have the same number of electrons in their outermost orbitals, which determines how they bond to other elements.

THE AWARD FOR THE MOST
METALLIC METAL GOES TO...

The periodic table also shows metallic behavior:
how easy it is for an element to lose an electron.
As you move to the right on the table, it becomes
harder to lose an electron, and the elements become
less metallic. As you move down the table, the
elements become more metallic. So the most
metallic metal, francium, is found in the bottom-left
corner of the table.

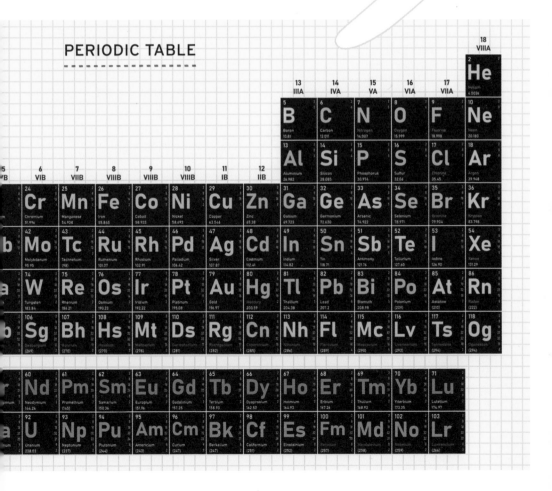

PERIODIC TABLE

EXPERIMENT:
EDIBLE IRON

Some chemical elements play important roles in keeping us healthy. A diet filled with lots of different fruits and vegetables is the best way to make sure we get enough of the elements we need, with each different vegetable helping in a slightly different way.

Bananas are rich in potassium (K), which helps nerves work and send signals to the brain. Lentils contain a lot of zinc (Zn), which helps the immune system, and spinach contains iron (Fe), needed for healthy blood.

It can be challenging to get enough healthy elements from some of the food on grocery store shelves. Breakfast cereal has always been a little controversial—often marketed to children and historically with very little nutritional value. In 1938, to help combat this, cereal companies started "fortifying" their products with vitamins and minerals, which meant adding some of the elements needed to stay healthy.

In this experiment, you will search for the iron hidden inside your food, and separate it out, in order to see it. Iron is one of the most abundant metals on Earth; in fact, it makes up over 30% of Earth's total mass. It is also useful. Humans have used it to make tools, weapons, and jewelry for thousands of years, and it has kept us healthy for much longer than that!

YOU WILL NEED:
- 150 g (5 oz.) of fortified cereal (for example, cornflakes)
- 2 ziplock bags
- Rolling pin
- Water
- Strong magnet (ideally neodymium)

WHAT TO DO:

1. Add the cereal to one of the ziplock bags and seal it. Use the rolling pin to crush the cereal into powder. Transfer the cereal powder to the second ziplock bag. The first bag may have holes caused by crushing the cereal.

2. Add water to the powdered cereal until the bag is half full. Squeeze some of the air from the bag and seal it.

3. Gently mix the bag's contents, allowing the cereal to dissolve, and leave it to stand for an hour. Take the magnet and place it on a table, and put the bag on top of it. Without moving where the bag is sitting, press gently on the bag to mix the contents around. You want all of the mixture to have a chance to pass close to the magnet.

4. When the mixture has been thoroughly moved around, carefully lift both the bag and the magnet together. Take care not to move the magnet away from the bag. If you look closely at the area in the bag next to the magnet, you may be able to see small dark specks—this is iron. If you move the magnet slowly, the specks should follow the magnet.

WHAT HAPPENS?

There is usually only about 1 g (³⁄₁₀₀ oz.) of iron per 10 kg (22 lb.) of cereal, so you won't find much, but it should be there. A daily intake of 18 mg of iron is enough to help keep your blood healthy!

IDENTIFYING THE IRON

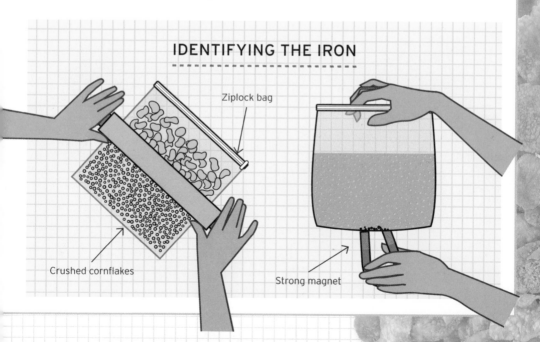

Ziplock bag

Crushed cornflakes

Strong magnet

LEARN ABOUT:
ELEMENT HUNTING

You've just uncovered the iron hidden in cereal. Now let's take a closer look at our relationship with other elements, especially those in food and drinks.

There are 118 different elements, and you are likely to find almost 60 of them in the human body (in tiny quantities). Humans only need around 20 elements to survive, but each has an important role to play.

ELEMENTARY, DEAR WATSON

It's time to do some more element detective work and see how many elements you can find in your home. To help in your quest, opposite is a list of the most common elements in the body.

- BORON (B)
- CALCIUM (Ca)
- CARBON (C)
- CHLORINE (Cl)
- CHROMIUM (Cr)
- COBALT (Co)
- COPPER (Cu)
- FLUORINE (F)
- HYDROGEN (H)
- IODINE (I)
- IRON (Fe)
- MAGNESIUM (Mg)
- MANGANESE (Mn)
- MOLYBDENUM (Mo)
- NITROGEN (N)
- OXYGEN (O)
- PHOSPHORUS (P)
- POTASSIUM (K)
- SELENIUM (Se)
- SILICON (Si)
- SODIUM (Na)
- SULFUR (S)
- TIN (Sn)
- VANADIUM (V)
- ZINC (Zn)

WHERE TO START?

Why not start your element hunt with a can of cola? Look at the ingredients—depending on the brand, it will probably read something like: carbonated water, sugar, colors, phosphoric acid, natural flavorings, and caffeine. While you might not be familiar with some of the terms, you can find two of our essential elements mentioned here, and we can expose another two if we rename water "dihydrogen monoxide."

VITAMINS AND MINERALS

In the previous section we looked at iron in cereal. Let's take a closer look at the other ingredients listed on the cereal box, paying close attention to the added vitamins and minerals. As with the cola can, you will probably find a few unfamiliar words, a mix of scientific and nonscientific names. For example, niacin might be called vitamin B3, or else its scientific name, pyridine-3-carboxylic acid. Most of the vitamins will contain nitrogen, bringing the total list of elements found to at least nine.

KEEP YOUR COOL

Raid your refrigerator and see what you can find. There is likely to be at least one easy element to spot here. Looking at the nutritional information on cheese, milk, or soy products should prove why dairy or dairy substitutes are so important for bones and teeth.

SPICE IT UP

You'll find two common elements in the spice cupboard. If in doubt, check the nutritional information on the salt container. You might find that your salt also contains ferrocyanide; the elements may be a little disguised here, but knowing that the Fe symbol for iron comes from the Latin *ferrum* is a clue. If you are using a low-sodium brand, you might find you can check potassium off your list too.

KEEP SMILING

Head to the bathroom and have a look at the ingredients in your toothpaste—you should find at least one tooth-strengthening element hidden here. While you're at it, try the medicine cabinet, which can be a treasure trove for the element hunter (ask an adult for help).

ELEMENTS

DISCOVER: CHEMISTRY CONNECTIONS

One thing almost every area of chemistry has in common is bonding. Chemistry is about the making, breaking, bending, and twisting of bonds, whether that is cosmic rays interacting with the gases in our atmosphere, or your body taking energy from the sugar in cola.

As we saw on pages 66–67, atoms comprise a dense nucleus of protons and neutrons surrounded by a cloud of electrons. To avoid the electrons colliding with each other, they are positioned in different orbitals. It is these electrons that determine the chemistry of the atom and what kind of bonds it will form.

THE VALENCE SHELL

There are many different electron orbitals, and not all follow a circular path, so to help understand, we group them into shells. The shells are like the layers of an onion, and they are filled from the inside out. The outermost shell, known as the valence shell, is involved in bonding.

When atoms bond, some want to gain more electrons, whereas others want to get rid of the ones they have—it all depends on how many electrons are in the valence shell. Let's say that the valence shell can hold eight electrons. The number of electrons we start with in the valence shell will determine how it reacts and bonds.

One or two electrons: The atom will want to get rid of these and will be left positively charged, because it now possesses a greater number of protons than electrons.

Between three and six electrons: The atom will want to share some electrons with other atoms to bring its total up to eight.

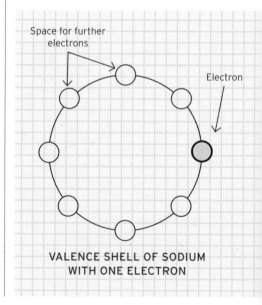

VALENCE SHELL OF SODIUM
WITH ONE ELECTRON

Seven electrons: The atom wants to fill its shell and will take an electron from another atom. This will leave it negatively charged because it now possesses a greater number of electrons than protons.

Eight electrons: The atom has a full shell and doesn't want to get rid of, share, or steal any electrons.

ALL OR NOTHING

Basically, atoms want either all or nothing in their valence shell. If they are less than half full, they lose their electrons; if they are around half full, they will share electrons; and if they are very close to filling the shell, they will steal what they need.

Atoms that lose electrons to become positively charged or steal electrons to become negatively charged are termed ions and can engage in ionic bonding. This happens with sodium and chlorine when they form table salt (sodium chloride). Atoms that prefer to share one or more electrons with each other form covalent bonds, such as when one carbon atom bonds to two oxygen atoms and creates a molecule of carbon dioxide.

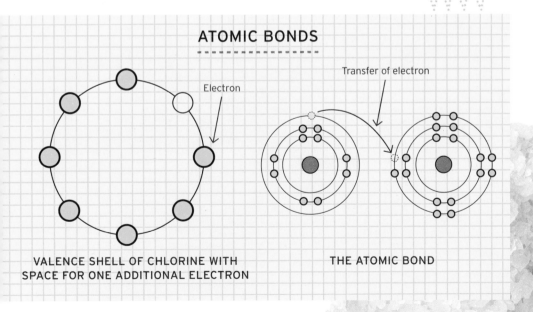

ATOMIC BONDS

Electron

Transfer of electron

VALENCE SHELL OF CHLORINE WITH
SPACE FOR ONE ADDITIONAL ELECTRON

THE ATOMIC BOND

LEARN ABOUT: PERIODIC SUCCESS

Test your knowledge of the elements and the periodic table by matching each element to its correct description.

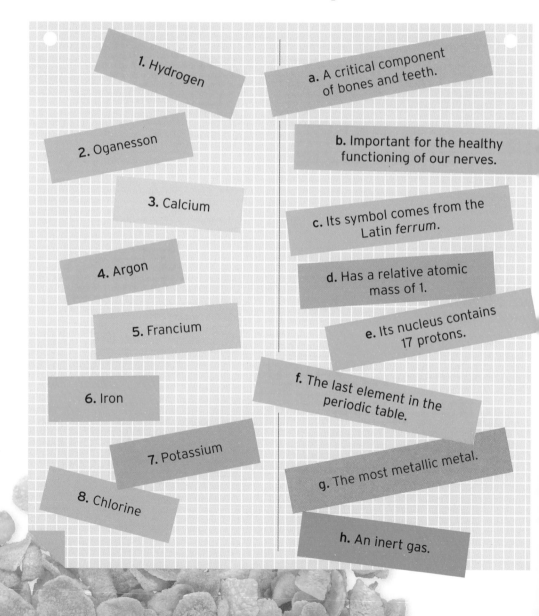

1. Hydrogen

2. Oganesson

3. Calcium

4. Argon

5. Francium

6. Iron

7. Potassium

8. Chlorine

a. A critical component of bones and teeth.

b. Important for the healthy functioning of our nerves.

c. Its symbol comes from the Latin *ferrum*.

d. Has a relative atomic mass of 1.

e. Its nucleus contains 17 protons.

f. The last element in the periodic table.

g. The most metallic metal.

h. An inert gas.

LEARN ABOUT: ELECTRON EXCHANGE

Combine your periodic table knowledge (see pages 68–69) with your bonding expertise and tackle these challenges all about reactions.

POP QUIZ: ELECTRONS

1. Each time a move is made one space to the right on the periodic table, one proton and (if the atom is neutral) one electron are added. How many electrons does element number 23 normally have? How many electrons does gold normally have?

2. As a general rule, when moving to the right on the periodic table, one electron is added to the outer orbital–the valence shell–of the atom. Moving to a new row on the periodic table means starting a new valence shell. This becomes more complex when moving into row four, so these questions will focus on elements in the first three rows.
a) How many electrons does oxygen have in its valence shell?
(Hint: You start at one every time you start a new row on the periodic table.)

b) Atoms like to fill their valence shells completely. How many electrons would oxygen need to gain to fill its shell?

3. Find argon on the periodic table.
a) How many electrons does argon have in its valence shell?
b) How many electrons can argon fit in its valence shell?
c) What kind of bonds would argon form?

4. a) Sodium is a very reactive metal. Looking at its electron configuration, why do you think this is?
b) Will sodium gain or lose electrons when it reacts?
c) Sodium reacts with chlorine to produce table salt, sodium chloride. What happens to sodium's outer electron during this reaction?

5. Which statement best describes covalent bonds?
a) Covalent bonds are formed by the gain or loss of electrons.
b) Covalent bonds are formed by the sharing of electrons.

6. One of the main constituents of cola is sugar. Sugar is a molecule composed of carbon, oxygen, and hydrogen. What kind of bond does carbon typically form– ionic or covalent?

DISCOVER:
MAGNIFICENT MOLECULES

In a bottle of cola there are normally at least five different types of atom: hydrogen, oxygen, carbon, nitrogen, and phosphorus. Thankfully, we don't buy our cola in an elemental state, or else we'd have a bottle of gas with some powders at the bottom. Instead, the elements are connected in various (and delicious) molecular combinations.

Molecules are made of atoms that are held together with chemical bonds. To help us picture how molecules work, we can think about atoms as letters and molecules as words. When we write, we connect letters to make words. In chemistry, we connect atoms to make molecules. In written language, the order in which the letters appear makes a huge difference—for example, "chemistry of cola" can easily be rearranged to make "frolicsome yacht." Two words can be composed of the same letters, but by using the letters in a different order, the words have different meanings. The same is true in the molecular world: two different molecules can contain the same number and type of atoms, but when rearranged in different ways, that same collection of atoms can produce two very different substances.

STRUCTURAL ISOMERS

Changing the order of the atoms but keeping the same number and type of atom creates a structural isomer. For example, butane has the molecular formula C_4H_{10}. Butane has a carbon backbone surrounded by hydrogens.

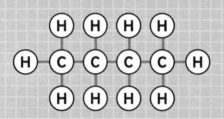

Not all molecules with the formula C_4H_{10} are butane, though; there are other ways in which these atoms can be arranged. Below you can see methylpropane, which also has the formula C_4H_{10}.

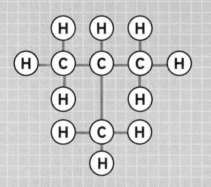

Instead of the chain of four carbon atoms, there is a chain of three, with one below. These molecules might be similar, but they behave differently. For example, butane will boil at 0°C (32°F), while methylpropane will boil at -12°C (10°F). The more atoms that have to be rearranged, the more pronounced the differences can become.

THE IMPORTANCE OF BONDS

It might seem like there could be endless combinations, but it all depends on how many bonds the atoms can form. Hydrogen only forms one bond, while carbon forms four. This means that all the hydrogens are bonded to just one other atom, while all the carbons are bonded to four others. The formula C_4H_{10} can only be arranged two ways: butane and methylpropane (see box, left).

Fortunately for us, the molecules in cola won't start switching around to form structural isomers. It takes reactive substances or lots of energy to break bonds and remake them, so the sugar and oils in your cola will stay sugar and oils.

DISCOVER:
SPEEDY SURFACES

Chapter 1 explored and explained a lot about ice: how it bonds, why it floats, and even how to create it instantaneously. There is one more question to consider: which cools a drink faster, crushed ice or ice cubes?

The answer is crushed ice, and the reason for that has to do with surface area, something that is very important in chemistry and physics. It influences rates of reactions and the properties of a material.

APPLES AND OXYGEN

For reactions to take place, there needs to be contact between the reactants, and when one of the reactants is a solid, the exposed surface area will influence the rate of the reaction. Think of an apple—thanks to its protective skin, an apple can stay fresh for a long time, but as soon as you take a bite out of it, the white flesh will quickly start to turn brown. The oxygen from the air starts a reaction, turning colorless substances in the apple to brown. The reaction won't start while the skin is intact because the apple's flesh is separated from the air. There will, however, be visible results quickly after the skin is broken.

With broken skin, the apple starts to decay. If we have taken a bite, then the whole bite area will start to react. If we cut the apple in half, both inside surfaces will start to react. If we chop the apple into slices, the reaction will start on every exposed surface. The higher the surface area exposed to the air, the faster the apple will decay.

UPPING THE SURFACE AREA

The reaction between apple flesh and oxygen in the air is the same, whether from a bite or chopping up the apple. But chopping the apple increases the surface area of exposed apple flesh that oxygen can reach. Sometimes increasing the surface area of a solid is as simple as just chopping it up to expose the inner surfaces.

Chemists have their own ways of increasing surface area. By creating the correct reaction conditions, chemists can produce spongelike materials that have holes, or pores, throughout the entire structure. These pores allow gases and liquids to penetrate deep into the solid and react throughout the entire material at once. This is used in water purification systems to clean water quickly.

Sliced apple will turn brown quickly.

DANGEROUS DUST

When large surface areas are involved, reactions can be dangerously fast. In 1981, an explosion occurred at a custard factory in Banbury, UK, after cornstarch powder and dust in the air caught fire. When firefighters put the fire out, water mixed with the unreacted cornstarch powder, and custard flowed down the streets.

BREAKING RECORDS

In 2018, scientists in Germany broke the world record for the material with the highest surface area. DUT-60 is a large cagelike molecule with lots of holes, and it has a surface area of 7800 m²/g. This means that a single gram of DUT-60 has a surface area equivalent to one and a half football fields.

LEARN ABOUT: DRAWING CHEMISTRY

Chapter 2 explained the different ways of writing out what is happening in a chemical reaction—using words or symbols. Another way of describing what is happening is to draw the molecules involved. Drawing molecules gives chemists a very easy and visual way of understanding a structure.

There are a few different ways to draw molecules, including as two-or three-dimensional drawings. We'll be sticking with 2-D drawings here, but it is worth remembering that molecules are 3-D, and in reality they will look different from our simple flat drawings.

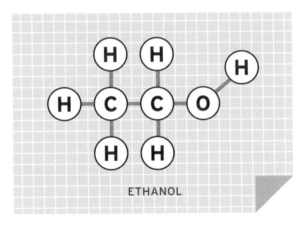

ETHANOL

BONDING ATOMS

Above is a drawing of an ethanol molecule, which is a form of alcohol. It has the formula C_2H_5OH. To draw it, we use the element symbols for the atoms, and lines to represent the bonds.

The bonds here show us where electrons are being shared. Atoms can form between one and four bonds; the number of bonds depends on the element. A hydrogen atom forms one bond, oxygen forms two, and carbon, four. In ethanol, you can see each atom has its preferred number of bonds, which makes this a stable molecule. The O and H at the right of the drawing should be regarded as a single OH unit, as in the formula.

MIX IT UP

We need to make sure that all the atoms have the correct number of bonds, but what happens if we change the order of the atoms? Remember that a molecule is made of atoms in the way that a word is made of letters, so changing the order of the atoms or the letters gives us something new.

CHALLENGE 1:

1. Ethanol has a structural isomer—a molecule with the same formula but different structure. Shuffle around the atoms in the ethanol molecule to form a new chemical, dimethyl ether. Remember that the number of bonds has to stay the same, so hydrogen forms only one bond, oxygen, two, and carbon, four.

(Hint: Try putting the oxygen atom in the middle.)

2. Methane is a flammable gas that in large quantities is harmful to the environment (cows are big producers of methane—it's a by-product of their digestive process). The formula for methane is CH_4; draw a methane molecule.

3. Ethane is another flammable gas and it has the formula C_2H_6. Draw this molecule. Does ethane have any structural isomers?

4. A pentane molecule—a liquid—is larger than a methane or ethane molecule. It has the formula C_5H_{12}. Draw pentane and work out how many structural isomers it has.

DOUBLE TROUBLE

Carbon likes to form four bonds, but not necessarily with four different atoms. If an atom wants to form more than one bond, it can form a double bond, such as in isoamyl acetate, which gives bananas their smell. In this molecule, the carbon and the oxygen have formed a double bond.

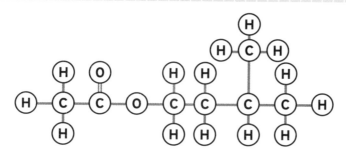

ISOAMYL ACETATE

CHALLENGE 2:

In a carbon dioxide molecule (CO_2), the two oxygen atoms form double bonds with the carbon atom. Draw the molecule.

LEARN ABOUT: SUSPECT SURFACES

Surface area and volume are extremely important for understanding chemical reactions, and every chemist must have a good grasp of the math involved. This section looks at the equations involved and puts them into practice.

POP QUIZ: VOLUMES AND SURFACE AREAS

Pages 80-81 looked at how important surface area is to chemical reactions. Depending on the surface areas involved, a reaction can be extremely slow or explosively fast. In order to understand and control reactions, scientists must have a good understanding of the math associated with surface area. In the same way, the ability to accurately calculate volume is also extremely important.

The equations on this page and on page 85 show how to calculate surface area and volume. Use them to help solve the problems on page 85. Radius (r) is the distance between the center of the circle and its outside edge; height (h) is the length of the object from bottom to top; and pi (π) is the distance around the outside of a circle (the circumference) divided by how wide the circle is (the diameter). For every circle π is 3.14.

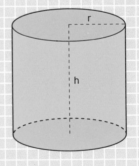

VOLUME OF A CYLINDER = $\pi r^2 h$

SURFACE AREA OF A CYLINDER = $2\pi rh + 2\pi r^2$

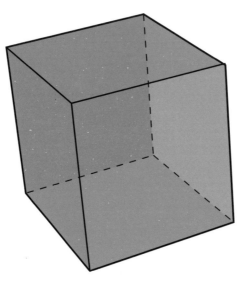

VOLUME OF A CUBE = a³

SURFACE AREA OF A CUBE = 6a²

1. The equation for the surface area of a cylinder is made of two parts added together. What do these parts each represent?

2. A can of cola is 330 ml and has a height of 11 cm.
a) What is the radius of the can? *(Hint: 1 ml is the same as 1 cm³.)*
b) What is the surface area of the can?

3. A block of ice cream has sides that are 25 cm long.
a) What is the volume of the cube?
b) What is the surface area of the cube?
c) You bore a cylindrical hole with a radius of 2.5 cm from one side of the cube to the other. What is the volume of this hole? And what is the new volume of the cube?

4. Surface area is especially important when considering how a liquid or gas might be quickly absorbed into another substance. Can you think of any items in your home which have a high surface area to make them absorbent?

EXPERIMENT: FIZZING FOUNTAIN

Pour a glass of cola and you'll see bubbles appear, grow, and float to the surface. If your cola is very fizzy, you might notice a steady stream of bubbles forming at the same spot in the glass.

Now dip the tip of your finger in the cola and look closely; you will probably see, and feel, that your fingertip is covered in tiny bubbles. Why do fingers cause bubbles? Because they are very rough surfaces.

The key to any chemical reaction or transformation is energy. Chemicals will always do things in the easiest way possible, and forming bubbles is hard work. The gas has to push against the liquid to form a bubble, and it needs help to do this. Bubbles will form at "nucleation sites"—defects, scratches, or rough areas where the disruption makes it slightly easier for a bubble to form, by reducing the energy required. A microscopic disruption to the smooth glass surface is all that is needed to allow the bubbles to form.

So, what happens when you introduce a very rough surface into a liquid with a lot of carbon dioxide dissolved in it? Let's find out.

YOU WILL NEED:
- 2 or 3 1-L bottles of cola
- Roll of chewy mints
- Sticky tape
- Lots of outdoor space and old clothes

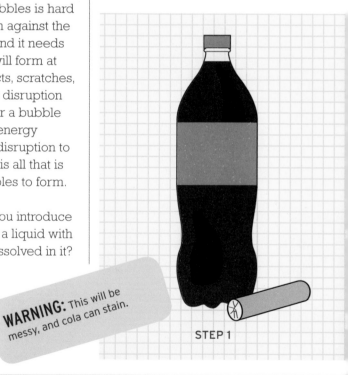

WARNING: This will be messy, and cola can stain.

STEP 1

WHAT TO DO:

1. Take one of the unopened bottles of cola and stand it in a flat, open outdoor space.

2. Take the roll of chewy mints and open the package along the side. Carefully reroll the paper back around the mints, but looser than before, and fix the wrapper back in place with some sticky tape. Leave one end of the wrapper tube open. The idea is that the mints should be able to fall quickly and freely out of the tube.

3. Ask an adult to help open the bottle of cola and to tip all the mints into the bottle very quickly. Stand back!

WHAT HAPPENS?

The mints have rough surfaces, and this means the dissolved carbon dioxide can nucleate all across the surfaces of the mints. The result: all of the gas will be released in a big whoosh! If you taste the cola afterward, you'll notice that it is already flat, whereas an open bottle can stay fizzy for hours.

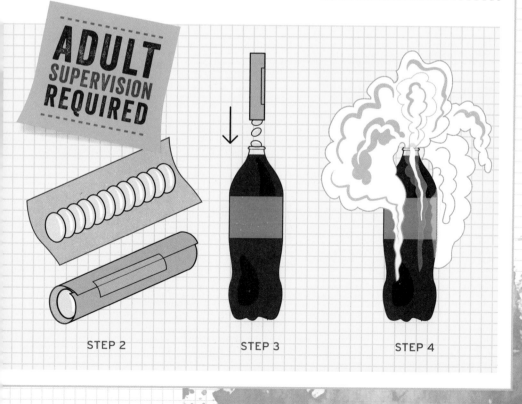

ADULT SUPERVISION REQUIRED

STEP 2   STEP 3   STEP 4

EXPERIMENT: DISCOVERING DIFFUSION

When a tea bag is put in hot water, the water turns a darker color. This is caused by diffusion, the movement of atoms or molecules from regions of high concentration to low concentration.

Particles in liquids and gases move past and bang into each other. This movement means that if you mix two liquids or gases, the boundaries between the two start to blur and the two mix. Using the tea bag example, the darker color produced by the interaction of the water with the tea bag spreads into all areas of the liquid. This continues until the whole mixture becomes the same color throughout.

Let's look at diffusion in action with this simple experiment.

YOU WILL NEED:
- Food coloring
- Tap water
- 3 large glasses

WHAT TO DO:

1. First prepare three glasses of water. Fill one glass with water and let it come to room temperature, then fill a second glass with water and add some ice cubes. After allowing the second glass to chill for a few minutes, remove the ice cubes and fill the third glass with hot water. Check that the volume of the water in each glass looks about the same.

2. Add a few drops of food coloring to each glass. Be careful to add the same amount to each glass (2–3 drops should be enough). Don't stir any of the glasses.

3. Start a timer. Watch the color diffuse through the three glasses, and record the time it takes for the water in each glass to become a uniform color. What effect does temperature have on the diffusion process?

EXTENDING THE EXPERIMENT

Diffusion can also be explored in these contexts:

a. Make and set a 2 cm (¾ in.) layer of clear gelatin in a shallow flat dish. Make two or three small dimples in the surface of the gelatin at large intervals and drop in some food coloring. You should see the color slowly spread through the gelatin in circular patterns.

b. Get a bag of different colored chocolate candies in a candy shell. Position the chocolates in a circle on a plate and carefully add water until they are all sitting in a shallow pool (but not submerged). The color will diffuse away from the chocolates, creating beautiful colored patterns in the water.

WHAT HAPPENS?

The warmer a substance is, the more thermal motion it has. The particles in the hot water are moving faster, and therefore diffusion should occur more quickly.

CHAPTER 4
FOOD CHEMISTRY

DISCOVER...

LEARN...

EXPERIMENT...

DISCOVER: SUGARY SCIENCE

If you ask someone to name all the ingredients in cola, sugar is likely to be at the top of the list. People have had a taste for sugar for thousands of years, and we are consuming more and more of it every year. But what is sugar? What is it used for? And why do we think it tastes so good?

When you think of sugar, chances are that you will picture the white crystalline powder. This is just one of sugar's many faces. Sugar is a carbohydrate—molecules made of carbon, hydrogen, and oxygen that play an important role in the body. Carbohydrates are used as fuel for cells; they are building materials; and they form the backbone of our DNA. The smallest of the carbohydrates are commonly referred to as sugars.

SWEET AND SIMPLE

The simplest of the sugars are called monosaccharides, which means "one sugar." They are small molecules with the general formula $C_nH_{2n}O_n$, where n is a number normally less than eight and C, H, and O are carbon, hydrogen, and oxygen. Two examples of monosaccharides are glucose and fructose, which you might have heard of—we will come back to them later.

If two monosaccharides bond together, they form a disaccharide ("two sugars"). Two common examples of disaccharides are sucrose (table sugar) and lactose (milk sugar). Sucrose is made of one glucose molecule and one fructose molecule bonded together. We can add more sugars to the disaccharide to create a longer chain—if we have between two and ten, we create an oligosaccharide ("few sugars"); if we add many more, we create a polysaccharide ("many sugars"). Oligosaccharides are used in our immune system and polysaccharides, like cellulose, are important building materials for plant cells.

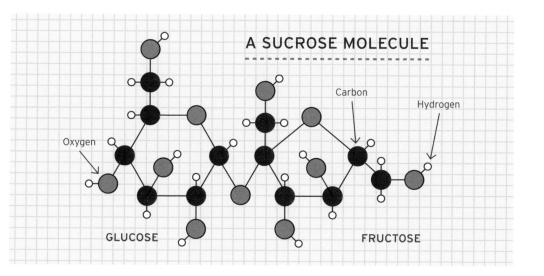

A SUCROSE MOLECULE

Carbon

Hydrogen

Oxygen

GLUCOSE

FRUCTOSE

Cola is likely to contain either a monosaccharide or a disaccharide. Most types of cola that contain sugar use corn syrup, which is a mix of fructose and disaccharide sucrose. The small fructose molecule can be absorbed through the intestine wall, meaning the sugar is available quickly for our body to use. But the sucrose in the cola needs to undergo another change before our body can use it—the body first breaks it down into its two monosaccharide parts, a fructose molecule and a glucose molecule. These two smaller molecules can then be absorbed by the gut and used as fuel for our cells.

The quick absorption and instant energy of sugar is partly why humans have learned to love the taste, but this is also leading to problems. Sugar gives us energy, but not much else. It contains no protein, vitamins, fiber, or fat—all of which our bodies need—so limiting the amount of energy we get from simple sugars, such as fructose, glucose, and sucrose, makes sense.

A cup of blueberries contains around 15 g (½ oz.) of sugar, as well as other carbohydrates and fiber.

POLYSACCHARIDES are polymers. Discover more about the fascinating world of polymers on pages 136–137.

EXPERIMENT: ONE SPOONFUL OR TWO?

The United States Department of Agriculture (USDA) recommends that no more than 10% of energy intake is in the form of added sugars; reducing intake even further increases the likelihood of health benefits. This is because eating too much sugar can not only cause you to gain weight; it can also increase the risk of disease.

The number of calories (energy from food) that you need depends on your age, health, activity level, and many other factors. The word "Calorie," which represents 1,000 calories, can be seen on food labels. In this book, "Calories" has been shortened to "cal." Recommendations vary, but the values of 2,500 cal for adult males and 2,000 cal for adult females are often given as approximate guidelines, with growing children needing somewhere between these values. If 10% of an adult's energy intake was from simple sugars, this would mean 250 cal for men and 200 cal for women.

Given that there are around 390 cal per 100 g (3½ oz.) of sugar, men are encouraged to limit their consumption to 64 g (2¼ oz.) of sugar, and women to 51 g (1¾ oz.). One can of cola is likely to contain around 35 g (1¼ oz.) of sugar—over half a man's recommended limit.

But while you would expect regular cola to contain lots of sugar, other foods that you wouldn't think of as being particularly sweet can also contain sugar, sometimes quite a bit. Foods that you might think of as being fairly healthy can also contain quite a bit of sugar. Let's take a look at exactly how much sugar is in different foods and drinks.

YOU WILL NEED:

- Pencil and paper
- Selection of foods and drinks
- Small squares of aluminum foil
- Kitchen scales
- Friend or family member
- Bag of sugar

WHAT TO DO:

1. Sugar content is typically listed on food and drink packaging per 100 g (3½ oz.) or 100 ml (3½ fl. oz.). Record the relative sugar content of the foods you've selected.

2. Using small squares of aluminum foil as trays, weigh out the amount of sugar in a few different foods and drinks. Set them side by side to see just how much sugar might be hidden in different foods.

3. Ask a friend or family member to match the samples to the correct food and drinks. How accurate were they?

WHAT HAPPENS?

You may have been surprised to find how much sugar there actually is in many of these foods and drinks. Fruit juices such as apple juice and orange juice can contain around the same amount of sugar, per 100 ml, as regular cola. Low-fat foods such as yogurt also contain quite a bit of sugar, which is used to enhance the taste lost by removing the fat. Even savory foods such as canned chicken soup contain sugar. It's easy to exceed the USDA's recommended sugar intake, without realizing it!

DISCOVER: FROM CROPS TO COLA

In most cases, especially in the US, sugar comes from corn. While you might not see many lab coats in cornfields, chemists have a crucial role to play in keeping food safe, plentiful, and tasting good. There are important chemical considerations at work to ensure good harvests.

Before a seed is even planted in the ground, the chemist's job has begun. Making sure the right level of nutrients are in the ground is essential for intensive farming. With the global population expected to be over 9 billion people by 2050, our farmland is having to feed more people than ever before. To help with this, fertilizers are used, in particular to provide plants with enough nitrogen to grow healthily. Plants can't take nitrogen directly from the air, so chemists process nitrogen into a more reactive molecule, which can then be used to make fertilizer. This process, called the Haber process, accounts for over 1% of total global energy consumption.

AGRICULTURAL chemistry looks after our food production, ensuring there is plenty of safe food to eat.

DEVELOPING DETERRENTS

Chemists also help to keep crops safe from pests, often by taking inspiration from nature. To create pesticides, chemists look at the natural defenses that plants have, and then try to enhance and adapt these to protect crops. For example, some plants give off chemical signals if an insect starts to eat them. Chemists analyze these chemicals with the aim of reproducing them on a large agricultural scale. By using natural deterrents, it is hoped that more extreme pesticides, which can kill insects and animals, might not be required.

Assuming a crop has had enough nutrition to grow and has been protected from insects, it is still far from ready to use as an ingredient in cola. After the corn is harvested, it is first milled down to produce cornstarch. The cornstarch is then mixed with water and an enzyme that starts breaking the large cornstarch molecules into smaller molecules. A second enzyme is then added to convert these smaller molecules into sugars—the result is corn syrup.

Corn syrup, which is 42% fructose and 58% glucose, can be used for baking, but it's not ready for cola just yet. Another enzyme is added to make the syrup taste sweeter. In this process, some of the glucose is converted to fructose, which makes the drink taste even sweeter because our tongue registers fructose as twice as sweet as glucose.

SWEET SYRUP

Corn is initially processed into cornstarch before it becomes the high-fructose corn syrup that is used in cola.

DISCOVER: INDESTRUCTIBLE ENERGY

Whether a reaction happens or not, and how it happens, depends on energy. Chemical reactions make and break bonds, and in doing so they can absorb or release energy.

Energy is complex, and its simplest definition is the ability to do work, which means the application of a force over a distance. Energy has the ability to change something, usually resulting in movement. It is measured in joules.

Some of the main forms of energy are:
Magnetic: the energy that flips two magnets when the wrong sides are pointing at each other.
Thermal: the motion of the atoms and molecules inside a substance.
Chemical: the energy stored in chemical bonds.
Elastic potential: the ability for a stretched or distorted substance to return to normal, such as with a rubber band.
Electrical: the energy involved in the movement of charged particles.
Nuclear: the energy stored in the nucleus of the atom.
Gravitational potential: the energy that causes an elevated object to fall to earth.
Kinetic: movement energy.

This is not a complete list; people disagree on the exact number of different forms. For example, is sound a form of energy in itself, or is it simply the movement of particles and therefore the same as kinetic energy?

INDESTRUCTIBLE

One of the most amazing things about energy is that it can't be created or destroyed; it can only change from one form to another. Our daily lives involve energy changing forms thousands of times. If you take a trip in a car, you are converting chemical energy (from fuel) into kinetic energy. When driving up a hill, you give the car potential energy, which uses more fuel, while coming down the hill you convert that potential energy to kinetic energy, and you might even need to use the brakes.

Energy doesn't change from one form to another with 100% efficiency. Some of the energy is always lost to the atmosphere, usually through conversion into heat or sound. This is why car engines get hot. Not all the chemical energy is transferred to kinetic energy. Over 50% of the chemical energy that we put into a car can be lost as heat.

For chemists, the most important form of energy is chemical energy, the energy stored inside bonds. The chemical in cola best associated with energy is sugar, which our bodies break down to release energy.

DISCOVER: ENERGY DRINK

A can of sugary cola can give you a real energy boost. Sugar is packed full of easy-to-access chemical energy, which can be rapidly used by our bodies. But how much energy does it take to power a human?

When you have breakfast in the morning, you take in food that has chemical energy stored inside it. Your body then converts the chemical energy into movement, heat, and even a small amount of electricity, which carries nerve signals around the body.

If you look at the nutrition information on a can of cola, you'll see a measure of the sugar energy written in Calories (cal). But the Calories used on food packaging is a measurement that is rarely used by chemists. The standard unit for energy is called a joule, and a Calorie is 4,184 joules.

IF HUMANS were powered with batteries instead of food, we would need more than 1,000 AA batteries every single day to keep us powered!

Nutritional Facts
Per 1 can (355 ml)

Amount per serving

Calories 140

	% Daily Value
Fat 0 g	0 %
Sodium 40 mg	2 %
Carbohydrates 42 g	14 %
Sugars 42 g	
Protein 0 g	0 %

If someone asked you if 800 km (497 miles) is a long distance, you'd probably say that it was. Similarly, if someone asked you if 2,000°C (3,632°F) is hot, you would probably agree that it was. Energy measurements, however, tend to be less familiar. Is 50 joules a lot? 500? What about 5 million?

You could obtain 10 million joules by simply drinking 12 cans of cola each day, but your body would protest. Cola's calories are "empty," giving us no other nutrition in terms of protein, fat, vitamins, and minerals, which are all required to keep us healthy.

ENERGY INTENSIVE

An average bottle of cola contains about 210 cal, which means it has 878,640 joules of energy, or enough energy to lift 878,640 eggs up 1.5 m (1⅕ yd.). If we convert our daily energy requirement to joules, we get a whopping 10 million joules!

This might seem like a huge amount of energy, but we need that much because our bodies use energy inefficiently. Only a portion of the energy we produce from our food and drink is used to make us run, jump, and kick a ball—much is simply lost as heat. Think about how hot you get when you run and jump; this is a by-product of an inefficient energy transfer process.

ENERGETIC EGG

A small experiment, which can be done in the kitchen, will help in understanding what a joule of energy is. Place a large, deep baking tray and an egg on the floor. Pick the egg up and lift it 1.5 m (5 ft.) off the floor above the tray. An object (the egg) with a mass of 65 g (2⅓ oz.) held 1.5 m (5 ft.) above the ground has 1 joule of potential energy. Dropping the egg onto the tray will convert the 1 joule of potential energy into 1 joule of kinetic (movement) energy. Finally, the egg will crack, releasing most of its energy in the form of sound, but also a fraction of a joule as heat.

EXPERIMENT: DENSITY RAINBOW

Can water float on ... water? It might seem like a strange question, but the answer is yes. Water can be made denser by mixing it with other ingredients. In this experiment you will create a rainbow of sugar water, by mixing different densities of sugar solution.

As the experiment on pages 22–23 explained, density is defined as mass divided by unit volume. It is given the symbol ρ and the unit kilogram per cubic meter (kg/m^3). To work out the density of a substance, the following equation is used:

$$Density = \frac{mass}{volume}$$

$$\rho = \frac{m}{v}$$

At room temperature, the density of water is $0.998\ kg/m^3$, but that can be easily changed by adding sugar. When sugar is dissolved in water, the solution gains mass, but its volume isn't increased. By creating different densities of water, you should be able to float different concentrations on top of each other.

YOU WILL NEED:

- Tall glass
- Food coloring (at least one, but ideally five colors)
- Granulated sugar
- Water
- Kitchen scale
- 5 small glasses
- Spoon
- Paper towel

ADULT SUPERVISION REQUIRED

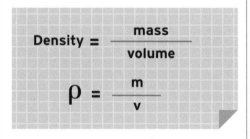

STEP 1

WHAT TO DO:

1. Add 50 ml (3 tbsp.) of tap water to each of your five small glasses, and add a few drops of food coloring to each. If you only have one color, add it only to glasses one, three, and five. If you have five colors, add a different color to each glass.

2. You won't be adding anything else to glass one, but add 15 g (1⅕ tbsp.) of sugar to glass two, 30 g (2⅔ tbsp.) to glass three, 45 g (3½ tbsp.) to glass four and 60 g (4⅔ tbsp.) to glass five.

3. Stir the glasses with the sugar. You should find that the sugar in glasses two and three dissolves quickly, while with four and five it will take longer. If after several minutes of stirring, solid sugar remains, heat the glasses in a microwave for short bursts (ten seconds at a time) to slightly warm the water. Sugar water will warm rapidly and can burn your skin, so be careful not to overheat the water, and be careful not to splash it on your skin. Ask an adult to help.

4. When you have dissolved as much sugar in glass five as possible, add it to the bottom of the tall glass.

5. Cut a strip of paper towel approximately 3 cm (1 in.) wide and 10 cm (4 in.) long. Dip one end in glass four and then place the wet towel against the inside of the glass; it should stick there.

6. Use the paper towel in glass four to guide the liquid gently into the tall glass by pouring the liquid slowly down the paper towel and into the glass, rather than pouring straight from one glass to another. The liquid from glass four should be less dense than the solution from glass five, and therefore it should float on top.

7. Repeat steps five and six with glasses three, two, and one (in that order).

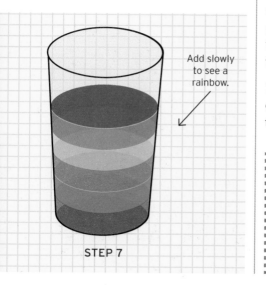

Add slowly to see a rainbow.

STEP 7

WHAT HAPPENS?

You should have a glass filled with layers of different colors and density, with water apparently floating on water.

EXPERIMENT: SINKING SUGAR

Can you tell the difference between cola and sugar-free cola? Most people probably can; the sugar content in cola makes it taste and feel slightly different in your mouth. But can you tell just by looking?

WHAT YOU WILL NEED:

- Can of regular cola
- Can of sugar-free cola
- Large basin or sink
- Water

WHAT TO DO:

1. Fill a basin up with water and put the can of regular cola in it. What happens? How buoyant is it?

2. Now add the sugar-free can to the water. Do you notice any difference?

WHAT HAPPENS?

You will likely find that the sugar-free cola floats, while the regular cola sinks. If it doesn't, you may have a soft drink that has a lower sugar content than others. Try to find a drink with the highest possible sugar content.

TESTING BUOYANCY

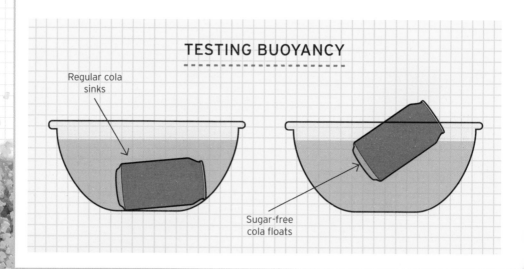

Regular cola sinks

Sugar-free cola floats

DEGREES OF SWEETNESS

A can of cola and a can of sugar-free cola have the same volume, but they have different densities. In sugar-free cola, the sugar is replaced with sweeteners, though not in the same amount. The artificial sweetener used in soft drinks is around 200 times sweeter than sugar and so a lot less is required. You'll find 35 g (1¼ oz.) of sugar in regular cola, and less than a gram of sweetener in the same volume of the sugar-free version.

VOLUME, MASS, AND DENSITY

Often, dissolving solids in liquids doesn't result in a change in volume; it will increase the mass and the density, instead. To understand this, picture a tall jar that you have filled to the brim with dry pasta shapes—the jar is now filled, but not entirely full. You could add rice to the pasta and it would fill the gaps between the pasta shapes.

To the pasta and the rice, you could then add salt. With some shaking, you would be able to fit large quantities of salt between the pasta and the rice. Even though the jar would be heavy and look very full, there would still be space to fit water in the jar.

When solids are dissolved in liquids, they start to fill the gaps between the particles in the liquid, and this is what allows sugar-free cola to have the same volume as regular coke but a lower density.

A FOLLOW-UP

As a second experiment, why not see just how much sugar your soft drink can actually hold? Lemon-lime soda would work best for this because it is colorless. Pour a can of lemon-lime soda into a glass and stir in sugar one spoonful at a time, counting as you go. At first the sugar will dissolve easily, but gradually more and more stirring will be required to dissolve the same amount. Eventually the sugar will no longer dissolve and will settle at the bottom, no matter how much you stir. When the sugar no longer dissolves, you have formed a saturated solution.

DISCOVER: SWEETER THAN SUGAR

Initially, sugar was added to cola in order to mask the bitter taste of the caffeine, but the sweet taste was a huge success and became one of cola's defining features. But as the consumption of cola grows, so do the problems associated with high sugar intake.

With each can of cola containing 35 g (7 tsp.) of sugar, cola takes a toll on teeth and waistlines. Soft drinks were revolutionized in 1981 when the first "diet cola" was introduced. This sweet but sugar-free drink changed drinking habits forever, and all because of some clever chemistry.

Fortunately, chemists have been experimenting with artificial sweeteners since not long after cola was first created in 1886.

ARTIFICIAL SWEETENERS

We expect cola to be sweet. If you tasted cola without sugar or sweeteners, the bitter concoction would be unrecognizable and pretty unpalatable. Keeping the distinctive taste, while removing the key ingredient, presented a huge challenge for cola manufacturers.

When we consume something sweet, cells in the tongue send signals to the brain to initiate a reward mechanism. It's important to remember that sugar isn't one chemical but a family of chemicals—sucrose, fructose, lactose, glucose, and more, which all activate these cells. The challenge in creating a sugar substitute was to find another chemical that activated the same cells, producing that same happy feeling.

SWEET CALORIES

Sugar-free cola uses aspartame instead of sugar, and the drink contains almost zero calories, so it's surprising to learn that aspartame contains more calories than sugar. Since aspartame is 200 times sweeter than sugar, only a fraction of the amount is needed.

SWEET SURPRISE

In 1879, after a late night in the lab, Russian chemist Constantin Fahlberg headed home for his dinner. Without stopping to wash his hands, he took a bite of bread and noticed it tasted incredibly sweet. He assumed it was a cake and rinsed his mouth out with water, but when he put his napkin to his lips he noticed an even stronger sweet taste. Fahlberg's hands were coated with his accidental discovery, saccharin, a sweetener more than 200 times sweeter than natural sugars.

Even sugar-free cola owes its success to luck. When chemist James Schlatter was working on a treatment for ulcers, he licked his finger before picking up a piece of paper and was shocked by a sweet taste. One of the compounds he had accidentally made was aspartame, which would be used to sweeten sugar-free drinks around the world.

Despite what these lucky breaks suggest, you should never put your fingers near your mouth while working in the lab!

REGULAR COLA

SUGAR-FREE COLA

THE PLANT Sclerochiton ilicifolius produces a compound, Monatin, which is 3,000 times sweeter than sugar!

DISCOVER: OIL, ESSENTIALLY

Simply mixing cola's ingredients, including sugar, acid, water, and caffeine, is not enough to produce the distinctive cola taste. There are also amazing molecules which give flavors and smells to cola and to the world around us—the essential oils.

Essential oils sound like they are pretty important, but unlike essential vitamins and minerals we don't need them to survive. Instead, "essential oil" refers to the strong-smelling liquid that can be extracted from many plants, including lavender, orange, cinnamon, and nutmeg. The liquid contains the fragrance, or essence, of that plant.

We've been using essential oils to create appealing smells and delicious tastes for hundreds of years, and there are a huge number to choose from—more than 150 have been approved for use in the food industry. In cola, you're likely to find cinnamon, vanilla, orange, nutmeg, and more. The biggest companies keep their individual blend a closely guarded secret.

VOLATILE AROMAS

These wonderful smelling chemicals are known as volatile aromatic compounds, a complex description. To a chemist, volatile means "easy to evaporate." Being volatile is useful for a chemical that is used to make pleasant smells; it means it will quickly and easily become a gas and therefore can be detected by your nose. A compound is also fairly simple to explain: it is something

VANILLA

made of more than one element, in this case mainly carbon, hydrogen, and oxygen. The word "aromatic" is a little more complex.

With all the talk of smell, you'd be forgiven for thinking that an aromatic compound is one that has a distinctive smell, but this isn't necessarily true in the world of chemistry. The word "aromatic" has a very particular meaning in chemistry. It refers to carbon-based molecules that have a ringlike structure with a distinctive electron configuration.

Below you can see two of the essential oil components found in cola—cinnamaldehyde (from cinnamon oil) and vanillin (from vanilla oil). Both molecules have a hexagonal ring with electrons shared around it, called the aromatic ring.

It isn't clear how the word "aromatic" came to describe a structural feature found in molecules. Some aromatic compounds smell nice, some smell bad, and others have no odor at all. It is assumed that the structural feature was first discovered in strong-smelling compounds, like essential oils, and therefore chemists wrongly assumed that these compounds always had an important role to play in smell.

Oils don't mix well in water, but very little is needed—just 0.5% of cola is comprised of essential oil flavors. Yet without them we'd just be drinking sugar water. There is a wonderfully complex world of volatile molecules that are working together to create that distinctive flavor.

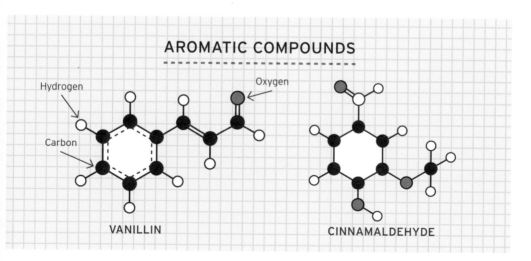

AROMATIC COMPOUNDS

Hydrogen

Oxygen

Carbon

VANILLIN

CINNAMALDEHYDE

DISCOVER: CAFFEINE KICK

Cola was originally developed as a restorative tonic, used to help people struggling with addiction to powerful drugs. To help produce this effect, cola's creator added caffeine to the mix.

Caffeine (chemical name: 1,3,7-Trimethylpurine-2,6-dione) is a molecule made of eight carbon atoms, ten hydrogen atoms, four nitrogen atoms, and two oxygen atoms. It is the world's most widely used psychoactive (mood- and behavior-altering) drug, with 85% of Americans consuming it every day. The reason it is classified as a psychoactive drug is because of the effect it has on our bodies—it can make us feel more awake and even improve our athletic performance.

In cola, the caffeine originally came from the kola nut, but it is possible to find caffeine in about 60 plants around the world, where it acts as a deterrent to pests. Insects that might eat the plants are put off by the bitter-tasting caffeine. Some of the highest concentrations of caffeine are found in coffee and tea. When ground coffee beans or tea leaves are brewed in water, they produce a caffeine-rich drink.

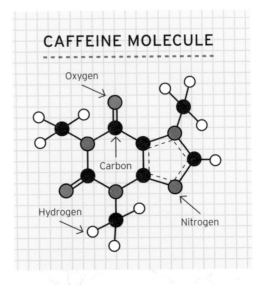

CAFFEINE MOLECULE

Oxygen

Carbon

Hydrogen

Nitrogen

WHAT HAPPENS WHEN A SPIDER DRINKS COFFEE?

A series of experiments were carried out in the 1940s and repeated in 1984 in which spiders were fed caffeine and then their web building was monitored. It was discovered that spiders couldn't build webs as well after ingesting caffeine; the webs became erratic and had large holes.

STAYING AWAKE

Typically, about an hour after drinking coffee, tea, or cola, the caffeine starts to have a noticeable effect and people feel more alert. This happens because caffeine has an effect on another chemical, adenine, which has a role to play in making us feel drowsy. Our bodies naturally produce adenine and it gradually builds up in our nerves while we are awake—the longer we are awake, the more adenine our nerves have. When our bodies detect increased levels of adenine, a reaction is started that makes us drowsy.

COFFEE BEANS

Caffeine interferes with the body's ability to detect the adenine and the drowsy reaction isn't started.

Caffeine doesn't just give people a morning boost; it is also in a medication that is given to premature babies who are having breathing problems, and has been found to increase the effectiveness of some over-the-counter medication such as aspirin. But it's not all good news—caffeine also increases blood pressure, can delay sleep, and may cause anxiety in some people. It can also become addictive—if we drink too much coffee or cola our body can start to expect it, and may experience minor withdrawal symptoms if we miss our daily dose.

OVER TIME our bodies adapt to caffeine intake and we need more and more for it to have the same effect.

DISCOVER: CRYSTAL CLEAR

Different things may spring to mind when you think of the word crystal: sparkling gemstones, fancy drinking glasses, or maybe even magic. Crystals can be found in smaller, everyday forms: tiny grains of sugar and salt have a microscopic beauty that is only found in the world of the crystal.

Understanding, growing, and controlling crystals is an important area for many science disciplines. After all, we eat them, wear them, and build many things out of them. So, what is a crystal? To understand this, we need to get inside the structure of the material and look at the atoms, ions, and molecules.

Amethyst specimen

REPEATING PATTERNS

Pages 12–13 looked at solids, liquids, and gases and explained that solids are made of atoms, molecules, or ions held closely together. Crystals are a special branch of these solids. In a crystal, the atoms, molecules, or ions form a regular repeating pattern. Imagine a chessboard: black, white, black, white, black, white—whether you go left or right, you know which color will be coming next. A crystal has a regular pattern like this, but in three dimensions.

Table salt is sodium chloride, which has the formula $NaCl$, from which we can see that there is one chlorine for every sodium atom. If we make our 2-D chessboard then we have a pattern of sodium, chlorine, sodium, chlorine.

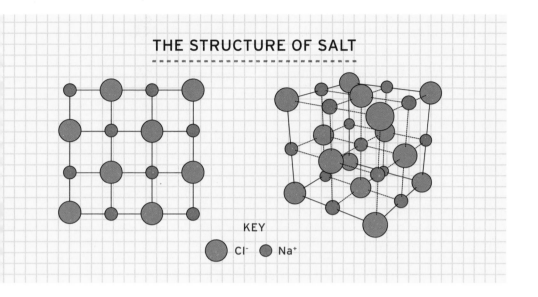

THE STRUCTURE OF SALT

KEY

Cl⁻   Na⁺

The diagram above shows eight sodium ions and eight chloride ions—one chlorine for every sodium, which is correct. But crystals are not simply flat, sheetlike structures, so layers need to be added above and below. The formula needs to work in all three dimensions, so there must be a chlorine ion above and below every sodium ion, and vice versa.

This is a salt crystal! A crystal with just a few dozen atoms would be tiny, just a fraction of the size of a cell in your body; in reality, salt crystals are much bigger. A single grain of salt contains over one quintillion (1,000,000,000,000,000,000) ions. This means our lattice (the repeating "chessboard" arrangement in the crystal) extends for billions and billions of ions in every direction. No matter how big the crystal, you know that if you start at a sodium ion and go up, down, left, right, forward, or back, you will find a chlorine ion.

SUGAR

Sugar is a complex molecule and the crystal looks quite different from the structure of sodium chloride. Sugar molecules don't fit neatly together, and their repeating pattern doesn't stretch as far as it does in salt. This means that sugar tends to form very tiny crystals. A single grain might contain millions of tiny crystals known as crystallites.

EXPERIMENT: GROWING CRYSTALS

When one of cola's main ingredients, sugar, is dried, it takes the form of a white crystalline solid. The size of the crystals depends on the conditions in which they are grown.

Try this experiment to see how large a crystal you can grow.

YOU WILL NEED:

- Wooden skewer
- Small pan
- 500 g (1 lb.) of sugar
- 250 ml (8½ fl. oz.) of water
- Tall glass
- Piece of stiff cardboard or a clothes peg
- Sticky tack
- Food coloring

ADULT SUPERVISION REQUIRED

WHAT TO DO:

1. Ask an adult to help simmer the water in the pan and slowly add the sugar in small 20–60 g (1–2 oz.) portions, stirring the water carefully to make sure the sugar dissolves. Be careful: hot sugar can cause burns!

2. Continue adding sugar until the water turns cloudy and no more sugar will dissolve. This is what is called a supersaturated solution—the water now has more sugar dissolved in it than it would tolerate at room temperature.

3. Remove the pan from the heat, stir in a few drops of food coloring, and allow it to cool.

4. Take a piece of thick cardboard and cut out a circle with a diameter that is at least 2 cm (¾ in.) bigger than the top of the glass. Use the skewer to poke a small hole in the middle of the cardboard.

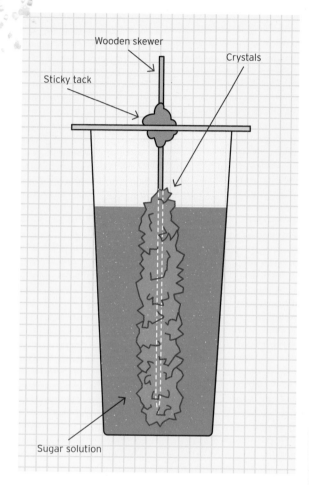

Wooden skewer

Crystals

Sticky tack

Sugar solution

5. Wet half of the wooden skewer and coat it in some of the leftover sugar. Feed the clean end of the skewer into the hole in the cardboard until the sugared end is 2 cm shorter than the inside of the glass. Secure the skewer in place using sticky tack on both sides.

6. Pour the sugar solution into the glass and place the sugared end of the skewer in the liquid, with the cardboard holding the skewer such that the sugared end is suspended at the center of the glass.

7. Store the glass in a place where you can leave it for up to a week.

8. After 2–3 days, you should begin to see crystals forming on the skewer. You might notice extra crystals growing on the surface. If you can, it is best to remove these.

9. When you feel the crystals are large enough, you can remove the skewer and examine it more closely.

WHAT HAPPENS?

The excess sugar wants to return to its crystal state, and the easiest place for a crystal to grow is on top of another crystal. This means that crystals can grow and grow!

DISCOVER: COLA'S SECRETS

The cola market is a competitive and potentially lucrative business, with cola sales bringing in over $14 billion in the US alone. Beverage companies want to create the perfect cola, but the largest brands keep their formulas a secret.

Two years after Dr. Pemberton created cola, he found himself struggling financially, and his health was failing. This led him in 1888 to sell his most precious asset—the original recipe for cola. Ever since, people have tried to uncover the secrets of the world's most popular soft drink. The air of mystery is one the leading brands have been keen to cultivate, and many myths and legends have been born.

Dr. Pemberton

MYSTERIOUS FLAVORS

You might think the mystery is easily solved with the ingredients listed on the side of the bottle in order of weight, but there is a big problem: "natural flavors." In the food standards world, a flavor doesn't count as an ingredient and therefore doesn't need to be listed on the container, as long as that flavoring is considered to be safe for consumption. This means that the unique taste of cola has remained much of a mystery.

The largest manufacturer of cola is so careful with its secret recipe that it keeps the ingredients hidden from its own employees. At a cola factory, you will see water mixed with nine different ingredients, which are labeled with numbers, not names. The first few are well known: sugar, color, caffeine, and phosphoric acid. The rest, however, are a bit of a mystery. The most mysterious of the ingredients is "7x," although it is believed to be a mix of essential oils to add flavor.

GAS CHROMATOGRAPHY

A mystery label isn't enough to fool a chemist, however. There are lots of techniques for unlocking the secrets of a mysterious liquid. One technique that has been used to explore cola is gas chromatography. Chromatography experiments can be carried out in the classroom, and they cover a range of different techniques, all of which are about separating chemicals.

In gas chromatography, a tiny amount of sample is injected into a long coiled tube, which has an inert gas (see pages 68–69) flowing through it. The gas and the chemicals travel through the tube and are detected at the other end. The length of time a particular chemical takes to pass through the coil depends on its size and chemistry. This means that each chemical in the mix is separated out.

When chemists looked at the top three brands of cola, they found 58 different aroma chemicals. Most of these smells and flavors arise from the essential oils (see pages 108–109). Some of the smells were expected, such as citrus, and others were bizarre, such as p-Cresol, which smells like pigs!

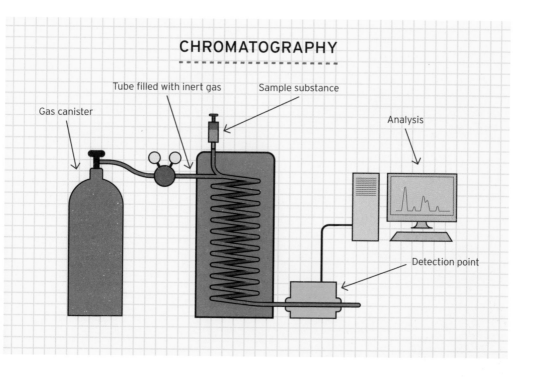

CHROMATOGRAPHY

Tube filled with inert gas   Sample substance

Gas canister

Analysis

Detection point

COLORFUL CANDY

The term "chromatography" covers a range of techniques. As well as gas chromatography, there is also liquid chromatography and thin-layer chromatography, plus several others. These all separate complex samples into their chemical components by passing them through a substance. The various chemicals in the sample travel through the substance at different speeds, resulting in their separation.

The chromatography techniques differ in exactly how they separate the chemicals and what kinds of samples they can be used to analyze. Gas chromatography works with gaseous samples, which are passed through a tube containing an inert gas. Liquid chromatography works with liquid samples, which are passed through a column filled with lots of tiny particles or a solid material with lots of tiny pores. Thin-layer chromatography also works with liquid samples, but separates them on a flat surface covered in a thin layer of gel.

There is also paper chromatography, used in this experiment to separate the dyes in colorful candies.

YOU WILL NEED:
- Candies that come in different colors, such as jelly beans
- Coffee filters (one per color of your candy)
- Scissors
- Small glass jar
- Cup of water
- Pitcher of water
- Pencil
- Sticky tape

WHAT TO DO:
1. Cut the coffee filters into thin strips, each around 2.5 cm (1 in.) wide and slightly longer than the height of the jar.

2. Dip one of the candies into the cup of water. Shake off any excess water and then press the candy down onto one of the filter strips, around 2.5 cm from one end of the strip, until some of the candy's color has rubbed off onto the paper.

3. Fold the opposite end of the filter strip over the pencil and secure it with the sticky tape.

4. Lay the pencil across the top of the glass jar so that the strip hangs inside the jar with the end not quite reaching the bottom.

5. Pour water from the pitcher into the glass jar so that it just touches the bottom of the filter strip, but doesn't reach the colored mark on the paper.

6. Watch for several minutes. The water will soak into the filter strip and travel up it. When the water reaches the splodge, it will carry the dye molecules in the color mark along with it, forming bands of different colors. Adding a small amount of salt to the water can aid this process.

7. When the water reaches the top of the filter paper, remove it from the jar and leave it to dry.

WHAT HAPPENS?

The colorful coatings on candies often contain several different dyes, which are soluble in water. Placing the candies in water releases these soluble dyes from the coatings, allowing them to be transferred to the filter paper to form a colored mark. When the water soaking into the filter paper reaches the colored mark, it again releases the dyes, but this time from the filter paper, and carries them along with it as it travels up the filter paper. However, different dyes will differ in their solubility, meaning how readily they dissolve in water, and in their tendency to stick to the filter paper. Those dyes with a high solubility and weak tendency to stick to the paper will travel faster and farther than those with a low solubility and strong tendency to stick to the paper. This results in the formation of different colored bands.

8. Repeat steps 2 to 7 with different color candies. Do the different color candies produce different numbers and colors of bands?

CHAPTER 5
LOOKING CLOSELY AT COLA

DISCOVER...

LEARN...

EXPERIMENT...

EXPERIMENT: COLORLESS COLA

Clear cola was invented in the 1940s, and in this experiment you will produce your own colorless cola—though you may not want to drink it!

In the 1940s and the 1990s, cola manufacturers tried introducing colorless cola to the world, but it proved unpopular with consumers. It was reportedly first produced after the Second World War to allow a Russian military leader to enjoy cola without being seen to like the distinctively "American" beverage.

YOU WILL NEED:

- Can of cola
- 100 ml (3 fl. oz.) of milk
- 2 large glasses
- Coffee filter or paper towel
- Funnel

WHAT TO DO:

1. Pour the can of cola into a large glass. Note the color and how much light you can see though the liquid. Stir the glass to rid the cola of some of the fizz.

2. Add the milk to the glass and stir well.

3. After 5 minutes you should start to see the milk reacting with the cola and starting to form a solid.

4. Stir the mixture every 15 minutes for the next hour.

5. Fold the filter (or paper towel) into a cone shape and put it in the funnel. Once an hour has elapsed pour the mixture through the funnel and into a second glass.

If the experiment has worked, you should have a colorless liquid which smells just like cola!

WHAT HAPPENED?

The milk has become lumpy; this is a process known as curdling and is the same thing that happens when milk spoils or is used to make cheese. To understand this process, you need to know a little more about milk as a liquid. Milk is a colloid suspension, meaning there are tiny undissolved solids floating in the liquid. In fresh milk, there are proteins and fat suspended in the liquid. These proteins don't normally interact with each other, but if the pH is lowered (the solution becomes more acidic) then the proteins unravel and start to clump together.

Since cola contains phosphoric acid, adding it to milk quickly starts the curdling process, producing visible protein clumps. But this still doesn't explain why the cola becomes colorless. The loss of color arises from an interaction between the color additive in cola (E150d) and the milk curds. The color chemical has an electrical charge, and it is attracted and sticks to the protein curds. When the curds are filtered out, the color that has bonded to them is also removed.

The result is a colorless and slightly less acidic cola, but it looks a lot less refreshing. Taste it, at your own risk!

STEP 2

STEP 3

STEP 5

EXPERIMENT: CLEANING WITH COLA

It turns out that cola isn't just for quenching thirst or providing an energy boost. In the right set of circumstances, it can be a handy household cleaning product.

YOU WILL NEED:

- 3 small plastic containers or glasses
- Distilled (white) vinegar
- Cola
- Water
- Paper towel
- 12 copper pennies

WHAT TO DO:

1. Note the appearance of the coins. Are they shiny? Are they dull? What color best describes them? If you have a camera, you could take a photo of them.

2. Place three coins into each of the three containers. In the first container, cover the coins with 1 cm (²/₅ in.) of water. Add 1 cm (²/₅ in.) of cola to the second container, and 1 cm (²/₅ in.) of vinegar to the third container.

3. Leave the coins overnight.

4. The next day remove one coin from each of the containers, rinse it with cold water, and dry it. Compare these coins with the three coins that were not in any liquid. Has the appearance changed? Are the coins shiny or dull? What color are they?

5. Remove the other two coins from each of the containers and, without rinsing or drying them, place them on paper towel and leave them for 24 hours. Dispose of the liquids down the drain with plenty of water.

6. Thoroughly wash your hands.

7. One day after removing all the coins, note the appearance for a third time.

8. Rinse all the coins in the vinegar, then wash and dry them. Wash your hands once more.

WHAT IS HAPPENING?

Copper coins are produced using a mix of different metals. Older copper coins will mainly be composed of copper, but these have become too expensive to produce, and newer coins may have a high steel content. Copper metal has a pinkish-orange color, but over time the copper in the coins reacts with the oxygen in the atmosphere to form copper oxide. The copper oxide coating tarnishes the coins, making them dull and darker in color.

Both vinegar and cola are acidic solutions—vinegar contains ethanoic acid, while cola contains phosphoric acid and carbonic acid. These acidic solutions attack the copper oxide layer on the coins. The ethanoic acid reacts with copper oxide to produce copper acetate and water, while the phosphoric acid will produce copper phosphate and water.

You should notice that the coins from the acidic solutions came out shiny and bright, while the ones in the water remained the same as the ones that were kept dry.

Finally, you might find that the coins from the acid that you didn't dry have gone slightly blue or green. That is because the copper ethanoate (green) and copper phosphate (blue), which were dissolved in the liquids, have been left behind when the liquid evaporated.

DISCOVER: A GASSY PROBLEM

Without the fizz, cola would be a very different drink experience. The tongue-tickling sensation of dozens of tiny bubbles adds to the experience of a soft drink. When you stop to think about it, soda is actually quite miraculous. It's a liquid that stores a gas, then releases it gradually. Here are the fizzy facts.

When William Brownrigg first managed to dissolve carbon dioxide (CO_2) in water back in 1740, little did he know what a phenomenon it would become. He hadn't discovered anything new—sparkling water occurs in some natural springs—but he had invented a way to do it artificially. Water has the ability to hold gases; this is what allows fish to breathe, their gills taking in the dissolved oxygen. In Brownrigg's case, he managed to increase the amount of carbon dioxide held in the water until it fizzed back out.

HENRY'S LAW

The fizzing is produced when the water has more carbon dioxide dissolved in it than it can handle, and it starts releasing it back into the atmosphere. The amount of gas a liquid can hold is governed by Henry's law, which was formulated by the English chemist William Henry in the 19th century. It tells us that the amount of gas that can be dissolved in a liquid is proportional, or directly linked, to the pressure above the surface of the liquid. The higher the pressure above the liquid, the more gas that can be dissolved.

We can see Henry's law in action when opening a bottle of cola. When the bottle is new and sealed, it is pressurized. This means that there is more gas squeezed into the space between the liquid and the cap than would be found in the same amount of space in the open air. At this point, no bubbles can be seen. The high pressure keeps the carbon dioxide dissolved in the liquid.

When opened, the pressure in the bottle is instantly reduced, and bubbles start to appear. We have lowered the gas pressure above the liquid and this means less carbon dioxide can be dissolved in the liquid. With the cola now holding more gas than it can cope with, it starts releasing it back into the atmosphere.

If the cap is left off the bottle, then most of the carbon dioxide will eventually escape. However, if the bottle is closed again, the bubble production slows because the bubbles which are popping on the surface are increasing the gas pressure inside the bottle. Eventually the dissolved gas and the pressure on the surface reach a new equilibrium, meaning they are balanced with each other. This means that more of the gas will stay trapped in the liquid and the fizz will last until the bottle is opened again.

A familiar fizz awaits.

BEWARE if you ever open a bottle of cola on a plane or on the top of Mount Everest— the low air pressure will mean the bubbles will be in even more of a hurry to escape!

DISCOVER: AN OCEAN OF COLA

Exploring the chemistry of cola tells us not just about the popular drink, but also helps us to understand the chemistry of the world around us.

Although the carbon dioxide that is added to cola is inert (not very reactive) in most cases, some of the molecules do react with the water and produce acid. When CO_2 is mixed with water (H_2O), we get this reaction:

$$CO_2(g) + H_2O(l) \rightleftharpoons H_2CO_3(aq)$$

Gaseous carbon dioxide plus liquid water reversibly produces an aqueous (dissolved in water) solution of carbonic acid in the cola.

The double arrow in the equation tells us that the reaction is in equilibrium. This doesn't mean that both sides have an equal amount of compounds or molecules. It simply means that they find a balance with each other. Imagine a seesaw with an adult and a child: the mass on either side is not equal, but the seesaw can be balanced if we move the adult closer to the center. Chemical reactions can be like this; they don't need to be equal on both sides to find a balance.

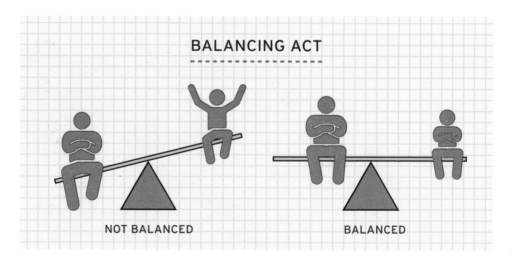

BALANCING ACT

NOT BALANCED   BALANCED

With carbon dioxide, the equilibrium is weighted heavily in favor of the carbon dioxide staying as a gas. Only about one in every 600 molecules of CO_2 will dissolve, producing the acid. Even though only a small amount dissolves to produce acid, it is still enough to alter the pH and taste of cola.

BURNING QUESTION?

Carbon dioxide dissolving in water to produce carbonic acid isn't a reaction confined to soft drink bottles; it is found the world over. Just as most bodies of water have enough oxygen dissolved in them to allow fish to breathe, most will also have detectable levels of carbon dioxide. Wherever carbon dioxide is dissolved, carbonic acid is produced.

Humans are producing CO_2 at record rates. The burning of fossil fuels, such as petroleum and natural gas, releases CO_2 into the atmosphere and eventually into the ocean. Around a quarter of our CO_2 emissions are absorbed by the oceans. The more CO_2 in a liquid, the more acidic it becomes, which means the pH of our oceans is falling.

Carbonic acid is weak and, although acid is never good for our teeth, humans are well adapted to cope with it. This isn't true for all life though—coral builds a skeleton of calcium carbonate (the same substance as in eggshell), which is very sensitive to acid attack. Mollusks that live in calcium carbonate shells, such as oysters, are also starting to feel the effect of ocean acidification, which is resulting in deformed shells.

LEARN ABOUT: KEEPING UP APPEARANCES

Test your knowledge of cola's appearance and its fizz.

POP QUIZ: COLA

1. What gives cola its brown appearance?
a) Caffeine
b) Phosphoric acid
c) E150d
d) Cinnamon oil

2. When cola is mixed with milk, the proteins begin to clump together in a process called curdling. What causes this?
a) The high carbon dioxide concentration
b) The high sugar content
c) An artificial color added to cola
d) Phosphoric acid

3. Copper coins can be cleaned with cola because the acid will remove the layer of tarnish on the coins. What is the dark layer that often covers old copper coins?
a) Natural oils from people's hands
b) Copper oxide produced by the reaction between copper and the atmosphere
c) Copper nitrate caused by exposure to air in moist atmospheres
d) Zinc cupric from reaction with other metals in the coin

4. Why is cola less effective at cleaning modern copper coins?
a) There is less copper in the newer coins.
b) Newer coins have an anticorrosion coating.
c) Older coins have been weakened by years of handling them.
d) All of the above

5. The amount of gas that can be dissolved in a liquid is directly proportional to the pressure of the gas above the surface. This is known as:
a) The McGlinchey partial gas law
b) Hooke's law
c) Brownian motion
d) Henry's law

6. When carbon dioxide is dissolved in water, what happens to the pH level of the solution?
a) It increases.
b) It decreases.
c) It remains unchanged.

7. If a chemical equation is in equilibrium, which of these statements is true?
a) No more of the product will be produced.
b) The reaction is complete.
c) There are equal amounts of product and reactant.
d) The rate of the forward reaction is the same as the reverse reaction, meaning they have found a balance and there won't be an increase or decrease in their concentrations.

8. Which of these suffer structural deformities if there is even a small increase in the volume of carbon dioxide dissolved in the ocean?
a) Plankton
b) Oysters
c) Lobsters
d) Sharks

9. Cola contains phosphoric and citric acids, and which other acid?
a) Carbonic
b) Ethanoic
c) Hydrochloric
d) Carboxylic

10. If you were able to open a can of cola deep under the ocean, where there is high pressure, what would you discover?
a) The volume of liquid would have greatly decreased.
b) It would seemingly have lost its fizz.
c) It would explode as soon as it was opened.
d) The color would have changed.

11. How much of our current carbon dioxide emissions are absorbed by the oceans?
a) 10%
b) 25%
c) 33%
d) 50%

12. Coral has a "skeleton" chemically similar to which other material?
a) Coffee mugs
b) Fingernails
c) Eggshells
d) Bone

13. Since the 1980s, countries that use copper coins have been reducing the amount of copper they use and replacing it with other metals. Why is this?
a) The copper corrodes too quickly.
b) The high electrical conductivity of copper makes it dangerous because we now routinely carry electrical devices in our pockets.
c) Copper reserves are running low.
d) Copper is more expensive, meaning the coin itself costs more than its value.

CHAPTER 6
COLA ON THE GO

DISCOVER...

LEARN...

EXPERIMENT...

DISCOVER: PLASTIC FANTASTIC?

When cola was first created in 1886, you couldn't drink it at home; it was sold by the glass at soda fountains. But the demand for cola grew, and manufacturers, scientists, and engineers had to work together to figure out a way for people to take fizz home.

Initially, cola was sold in glass bottles. While these are still used today, glass does pose some problems for the food industry, including that a glass bottle can smash, and it is quite heavy. In 1947, a revolutionary new product burst onto the scene—the plastic bottle. Today, around a million bottles are sold worldwide each minute!

Plastic is an amazing material. Although there are many different types of plastic, they are all, in general, lightweight, durable, strong, and inert. This makes them the perfect material for storage and packaging. However, there is a significant problem: plastics don't biodegrade easily. Biodegradation is the process whereby a material breaks down naturally over time. Plants and animals biodegrade after they die. Plastics, on the other hand, last for hundreds of years, littering the environment and polluting the oceans.

POLYMERS

Plastics are polymers. While the word "plastic" describes a material's ability to change shape when pressure is applied, the word "polymer" describes its molecular structure. "Polymer" comes from the Greek *poly*, meaning "many," and *méros*, meaning "part," and that's exactly what a polymer is—a material made up of many parts.

The "parts" of a polymer are called monomers. They are small units connected together in long chains. Imagine a string of holiday lights: each individual light is like a monomer, and each is connected to two other monomers in a chain.

GETTING IN A TANGLE

The holiday lights can help us understand some of the properties of a polymer too. If you take a set of holiday lights from storage, you might find they've become tangled up in a big ball. This happens to polymers, too. The long chains in a polymer get very easily twisted and tangled around each other, creating a strong material. The longer the chain, the more entangled and rigid the polymer is. If the chains were very short and didn't get tangled, the plastic would be much softer and melt at lower temperatures. Even shorter chains would result in the polymer behaving like a liquid.

When the ball of holiday lights is untangled, it becomes one long string of lights. It's possible to do something similar with polymers, too: as you stretch a polymer, the chains untangle and can eventually be lined up in fully extended rows. If you were to continue pulling, the chain would eventually break.

NOT all polymers are plastics but all plastics are polymers!

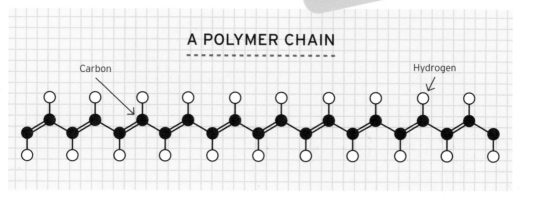

A POLYMER CHAIN

Carbon

Hydrogen

DISCOVER: A BRIEF HISTORY OF POLYMERS

Polymers have been around since life began, but human-made (synthetic) polymers are only about 100 years old. As with so many scientific discoveries, polymers were developed through first learning about and then adapting the natural world.

Humans initially used the natural polymers wool and cotton, but since relatively recently we've been able to produce synthetic polymers to suit our every need, using them to manufacture a wide range of items, from cola bottles to spacecraft parts.

One of the first big changes in polymer science came in 1820, when Scottish chemist Charles Macintosh invented waterproof fabric. His ingenious idea was to dissolve natural rubber with naphtha, a by-product of oil and tar. The naphtha allowed the rubber to be thinly sandwiched between two layers of fabric, creating a waterproof layer in the middle. This gave rise to the famous Macintosh raincoat, and also prompted scientists to begin mixing natural polymers with petrochemicals, which paved the way for synthetic polymers.

BILLIARDS WITH A BANG

Fast-forward a few decades to 1870 and polymers were about to solve another unusual problem. Before the 20th century, many billiard balls were made from ivory. With growing global awareness and elephant populations declining, and demand for billiard balls increasing, a company decided to launch a competition to find a material to replace ivory that could be produced in a factory. The company offered a $10,000 prize for anyone who could develop it.

One of the contenders for the prize was the American inventor John Wesley Hyatt. By experimenting with some natural polymers, Hyatt discovered a way of producing solid nitrocellulose, which behaves in a similar way to ivory. There was one downside to nitrocellulose, which Hyatt named celluloid—it is explosive! Although it was stable most of the time, the billiard balls were sometimes heard to sound a bang, like a gun going off. No one is sure if Hyatt ever got to collect his $10,000 prize, but he made it into the National Inventors Hall of Fame for his efforts.

Fortunately, today's polymers are a lot safer than those created in the 1800s. One of these is polyethylene terephthalate, known as PET, a lightweight, strong, and waterproof material invented in the 1940s that is now used to make hundreds of billions of bottles each year. The chemicals in PET bottles come almost entirely from oil, linking it to Macintosh's work 200 years ago. While these may seem perfect for cola, PET bottles last for over 400 years, so it's important that we recycle them.

NATURAL SOLUTIONS

Scientists are again looking to nature to find new materials, so that we might benefit from the properties of a polymer without the environmental cost. Chemists are trying out all sorts of ideas, from biodegradable polymers that can be composted, to worms that can digest long-lasting polymers.

DNA, which carries the genetic code of plants and animals, is a polymer!

DISCOVER: PLASTIC POLLUTION

The world has a problem with plastics. Over 8 million tonnes (8.8 million tons) of plastic finds its way into the oceans each year. Before attempting to solve this problem, it is necessary to understand the chemistry behind it. The problem of plastic pollution is stark: by 2050 the combined weight of plastic in our oceans could well match that of all the fish. As many as 90% of all seabirds have plastic in their digestive tracts.

The reason polymers have proved so popular is that they are lightweight, durable, and inert. The term "inert" means that the polymers will tend to remain unchanged, so unlikely to react, decompose, or form something new. Being inert is a great property for food packaging because it won't decompose when it touches food, and more importantly, it won't react with or contaminate what we are going to eat. The inert property of polymers only becomes a problem when we have finished using them.

BREAKING THE CHAIN

To understand why many polymers are inert, we need to think again about their structure. Pages 134–135 described polymers as long, chainlike molecules. To decompose the polymer, you need to break the chain, and that is very difficult to do because it has strong bonds and a long "backbone" of atoms. It also has other atoms bonded all along its length.

A simple polymer such as polyethylene has a carbon backbone with hydrogen atoms all along the sides. These

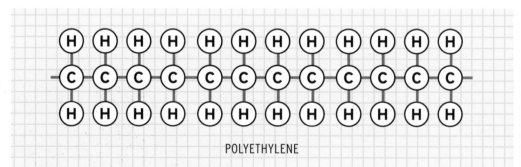

POLYETHYLENE

hydrogen atoms help protect those carbon-to-carbon bonds from attack.

Ultraviolet light emitted by the Sun can sometimes break the carbon-to-carbon bonds. When UV light strikes a polymer, it might be absorbed or reflected, and only some of the interactions with the right wavelength hitting the right spot will result in a reaction. When a reaction does take place, it will happen at a random position along the length of the chain. It might break a chain with 1,000 parts in half, leaving two chains of 500 parts, or it might just chop the end off.

Even when sunlight breaks bonds in a polymer, the polymer may remain largely unchanged. The broken polymer will still consist of long chains, and the sunlight would have to break hundreds or even thousands of bonds to degrade the plastic entirely.

In the ocean, it is more difficult for UV light to reach the polymers, and when it does they are likely to just be broken down into smaller and smaller pieces, eventually becoming microplastics and continuing to litter the oceans.

EXPERIMENT: SHRINKING THE PROBLEM

Polymers can be made into all shapes, sizes, and colors, from bright-red plastic bags and transparent cola bottles to opaque pipes. You've learned that many polymers can be stretched, causing the long-chain molecules to untangle. In this experiment, you will take a look at this process and discover how to recycle polymers.

You'll be investigating polystyrene, one of the most common polymers. You can find polystyrene in many different forms: the transparent and brittle cover of CD cases; the white foam "peanuts" used to protect boxed items; and disposable cutlery. You can identify polystyrene via the resin identification number 6 (see page 144), which is marked on packaging inside a triangle made of three arrows.

The focus of this experiment is the very thin and transparent kind of polystyrene. For the purposes of this experiment, the best kind is the transparent clamshell packs, which are often used to store cakes and salads.

YOU WILL NEED:

- **Thin sheets of transparent polystyrene (resin identification number 6)**
- **Coloring pens**
- **Oven**
- **Scissors**
- **Aluminum foil**
- **Baking sheet**
- **Oven mitt**

WHAT TO DO:

1. Find some clear polystyrene from the recycling (you can also purchase sheets online). Make sure it is not expanded polystyrene (packing peanuts), which is not transparent. Cut any textured edges off, leaving a clear, flat sheet.

2. Draw and color in some shapes on the polystyrene. You can also punch some holes to hang your creation. Once you have decorated some areas of the sheet, cut around your drawings so that you have several decorated polymer pieces.

3. Preheat the oven to 180°C (350°F).

4. Place the colored polymer shapes on a piece of aluminum foil on a baking sheet, and put them in the oven, on the lower rack. Ask an adult to help.

5. As the polymer pieces warm, they will twist and distort before starting to shrink and finally, becoming smaller, thicker pieces. Do not let the polymer burn.

6. When the pieces have stopped moving, the sheet is ready to be removed from the oven. Ask an adult to help. Use an oven mitt to remove the tray, and be careful not to touch the pieces as they will be very hot. Allow them to cool. As they do, they will set into their new shape.

WHAT IS HAPPENING?

The polystyrene is shrinking back to its original shape. When polystyrene is heated, the polymer chains begin to move around more easily. From initially being stretched straight, they become coiled and bent. Polystyrene is produced in sheets about 3 mm ($1/10$ in.) thick, which are then heated and stretched to be very thin, and used to produce food packing. By heating the polystyrene pieces, you're enabling them to return to their original thickness.

POLYSTYRENE ART

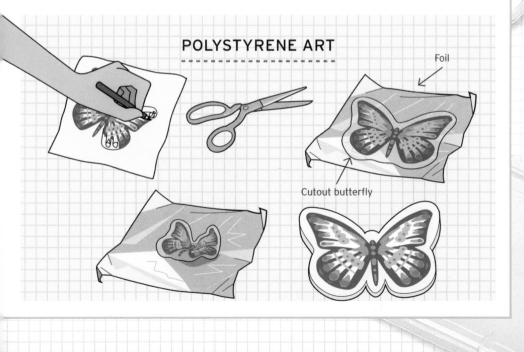

Foil

Cutout butterfly

DISCOVER: CAN IT

There are 180 billion drink cans produced each year. Unlike plastic bottles, which are made of long complex molecules, drink cans are, chemically speaking, quite simple: they are made almost entirely out of one material—aluminum.

Making the cans from aluminum works well because it is abundant, lightweight, and strong. That said, it takes a lot of chemistry know-how to produce the perfect drink can.

PRECIOUS METAL

If you had told someone in the late 19th century that in the future aluminum cans would be found in bars, supermarkets, and homes across the world, they would have laughed at you. You may as well have suggested that we build houses out of gold or build furniture from solid silver. Aluminum was once one of the precious metals, along with gold, silver, platinum, and palladium—this despite it being the most abundant metal in Earth's crust.

So why was aluminum worth so much? Although it is common in the ground, it is extremely rare in its pure elemental state. You might remember from page 67 that an element is a substance that contains just one type of atom. It is very rare to find pure aluminum in the ground—it has almost always reacted with silicon and oxygen to form a compound.

ALUMINUM

In the 1850s the French emperor Napoleon III had cutlery made from aluminum to show off his lavish wealth, although it was reserved for only his most favored guests (the less important diners had to tolerate gold and silver cutlery). Things were about to change quickly, however.

ALUMINUM ALCHEMY

In 1886, a French chemist and an American engineer developed the Hall–Héroult process, whereby an aluminum compound could be converted into pure aluminum. From that point on, the price of aluminum started dropping.

Three years later, an Austrian chemist developed the Bayer process. This allowed aluminum-rich rocks to be processed into a form that could be used in the Hall–Héroult process. The price of aluminum plummeted and demand grew. Thanks to those three scientists, we now produce over 60 million tonnes (66 million tons) of aluminum each year, using it for drink cans, aluminum foil, and aircraft.

ENERGY INTENSIVE

There is one big drawback to this marvelous refinement process: it requires lots of electricity to separate the aluminum from the other materials. In Australia, the aluminum refinement process accounts for 12% of the country's total electricity consumption! Fortunately, once we have the aluminum it can be reused time and time again by collecting and recycling—Switzerland manages to recycle over 90% of its aluminum cans every year.

GEMSTONES

If the conditions are correct, aluminum and oxygen can combine with other metals to produce rubies and sapphires. The color will depend on which other elements are included.

DISCOVER: RECYCLING

Once a bottle or a can of cola is empty, it will (hopefully) be bound for the recycling plant. While recycling has become routine in many parts of the world, it is still misunderstood and occasionally done incorrectly.

Maybe you've found yourself hovering over a choice of recycling containers, wondering where your trash belongs, or trying to decide what material you have and whether it can be recycled. It isn't always as simple as we'd like it to be.

SORTING THE TRASH

The key to good recycling is sorting. If you can gather all of the same kinds of material together, you can begin to reprocess it. Deciding which container to put things in is just the start of the sorting process.

As we've seen, aluminum cans can be easily recycled. Huge magnets sort the cans from the rest of the waste material, and they are sent to be melted down and reprocessed.

Polymers are far more complex, however. If you look at plastic packaging, somewhere on it you should be able to find a small triangle made of arrows with a number in the middle. This is a resin identification code.

The International Resin Identification Coding System gives each polymer a number between one and seven. Looking at these numbers will tell you what the object is made of. For example, most plastic cola bottles will be marked with the number one, which tells us that they are made of polyethylene terephthalate (PET). Objects made of PET are some of the easiest to be recycled, though they must first be separated from the rest of the trash and collected together.

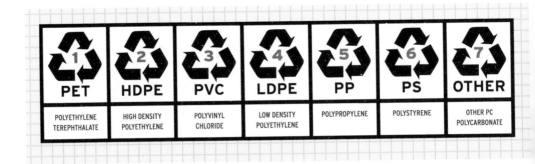

THE RECYCLING PLANT

When recycling arrives at a processing site, it will normally start its journey rumbling along a conveyor belt, where people will remove anything that is unlikely to be recyclable. Next, little jets of air blow lighter material like paper and cardboard into a separate channel, leaving plastic and metal behind. Magnets remove the aluminum and steel for separate processing. What's left is a conveyor belt filled with different polymers, and that's when things get really clever.

To sort one type of plastic from another, an infrared camera scans across the conveyor belt. It can instantly differentiate between the various types of polymers, and jets of air separate them into their different types. It takes a fraction of a second for the camera to identify the type of polymer and to push it into the correct stream. The camera has just one problem: it can't see black plastic. Therefore, many of the black microwave meal trays are destined to be missed and end in landfill.

With the polymers sorted, some, such as our cola bottle, will be recycled. (Others, such as number six, polystyrene, which is more difficult to reprocess, often end up in landfill.) The cola bottle will go on to be sorted by color, ground up into small pieces, and melted down. Sometimes the material becomes a bottle again, but often it will be used for other polymer-based products, such as carpets.

GREEN CREDENTIALS

Less than a quarter of plastic bottles used in the United States are recycled. Germany leads the way, with over half of all their trash recycled.

LEARN ABOUT: POLYMER PROBLEMS

We've **delved** into the world of polymers, and now it's time for you to **test** what you've learned.

POP QUIZ: POLYMERS

1. What property does the word "plastic" describe?
a) Unreactive and smooth material
b) Meltable
c) Can be shaped or bent
d) Packing material

2. What are the repeating units of a polymer called?
a) Elements
b) Styrenes
c) Isomers
d) Monomers

3. What are most synthetic polymers made from?
a) Modified tree sap
b) Petrochemicals
c) Sugars and starches
d) Acid and mineral reactions

4. Many polymers have a long carbon chain as their backbone, but which other element is also used to make polymer chains? *(Hint: Look at the trends in the periodic table on page 68.)*
a) Oxygen
b) Aluminum
c) Sulfur
d) Silicon

5. Which of these is not a polymeric substance?
a) Glass
b) Wool
c) Cotton
d) DNA

6. Select the correct end to this sentence: Polymers with shorter chain lengths typically . . .
a) occur only in nature.
b) melt at lower temperatures than long-chain polymers.
c) are tougher and stronger than polymers with longer chains.
d) have been artificially created.

7. Plastic drink bottles are made of:
a) polysaccharide
b) polyvinylchloride
c) polyethylene terephthalate
d) polytetrafluoroethylene

8. How many plastic bottles are bought every minute around the world?
a) 0.5 million
b) 1 million
c) 10 million
d) 100 million

9. By the year 2050, it is estimated that the plastic in our oceans will weigh more than what?
a) Three jumbo jets
b) The mass of rubbish produced by Sweden annually
c) All the Atlantic Ocean's population of humpback whales
d) The combined weight of all fish in the ocean

10. The global plastic problem has arisen primarily because:
a) Polymers break down easily in water.
b) Polymers alter the pH of seawater.
c) Polymers litter the ocean floor, which means less space for plants and sea life.
d) Polymers don't easily biodegrade.

LEARN ABOUT: RECYCLE AND REUSE

Test your recycling knowledge by tackling these problems.

POP QUIZ: RECYCLING

1. How are most different types of plastic separated from one another in a recycling plant?
a) By hand
b) By weight
c) With targeted jets of air
d) Via centrifugal force

2. What percentage of seabirds around the world are estimated to have plastic in their digestive system?
a) 20%
b) 46%
c) 73%
d) 90%

3. Polymers are stamped with a number inside a triangle made of arrows. What does this tell us?
a) Whether it can be recycled
b) What type of polymer it is made from
c) The length of time it will take to biodegrade
d) The purity of the polymer

4. Which of these countries recycles or composts more of their waste than any other?
a) The United Kingdom
b) The Netherlands
c) Germany
d) The United States of America

5. Approximately what percentage of the plastic bottles used in the US end up in recycling?
a) 10%
b) 25%
c) 40%
d) 50%

6. Which of these is regularly made from recycled plastic drink bottles?
a) Packing peanuts
b) Clamshell packaging
c) Carpets
d) Shoe soles

7. How many years does it take for a discarded plastic drink bottle to degrade?
a) 200-400 years
b) 400-600 years
c) 600-800 years
d) It will not degrade

8. Polymers can stay in animals for many years. Why is this?
a) They bond to the soft tissues in the digestive tract.
b) The pieces tend to be too small to be noticed inside the body.
c) Polymers are recognized as a natural part of the body.
d) The inert polymers are not able to be broken down by the body.

DISCOVER: BOTTLE IT

So far we've looked at plastic cola bottles and aluminum cans. To finish, let's look at where it all began: the glass bottle. Like plastic, glass is an amazing material that is used in countless ways in our daily life. It may even be set for a comeback as we try to reduce our plastic waste.

Glass is one of the most important materials in our daily lives; from windows to cola glasses, you will come across different forms of glass every day of your life. It is also a material that humans have used since the dawn of civilization. While we've been creating glass products for only the past 4,000 years, there is evidence of natural glass (caused by lightning strikes and volcanoes) being used as far back as the Stone Age.

Glass is composed mainly of silicon dioxide (SO_2), which you can find in abundance as sand. Pure silicon dioxide will melt at 1,650°C (3,000°F), and this results in a transparent solid that is very resistant to changes in temperature. We call this form of glass fused quartz. In theory, fused quartz could be used to produce bottles for cola, but working with temperatures over 1,650°C is very difficult and expensive. This is above the melting point of many metals—for example, iron melts at 1,535°C (2,795°F).

BRINGING OUT THE BAD CRYSTAL

Maybe somewhere in the house you have some "crystal" glasses kept for special occasions. Don't bring these out if you want to impress a chemist! The crystal in the cupboard isn't a crystal; in fact, it is the very opposite. A crystal is a material with a very ordered and predictable structure; glass, however, is a material that has no clear order or arrangement.

SODA AND LIME

In order to produce glass that has the strength and transparency of fused quartz but is easier to work with, various chemicals are added to the mix. The most common glass in daily use is known as soda lime glass. While SO_2 is still the main ingredient in soda lime glass, you will also find sodium carbonate (Na_2CO_3), which lowers the melting temperature, calcium oxide (CaO), which makes the glass less reactive, and magnesium oxide (MgO), which makes it hardwearing. This resulting mixture produces a glass that can be softened and worked at temperatures around 700°C (1,292°F).

The lower working temperature of soda lime glass makes it much easier and cheaper to process than fused quartz. Soda lime glass lends itself well to the recycling industry too because bottles can be melted and reprocessed. Whereas plastic bottles often contain a variety of pigments to produce a limitless variety of colors, glass tends to be clear, green, or brown. This also helps with the recycling process because it is easy to categorize the materials before they are melted down, meaning a pure recycled product can be easily obtained.

We've been drinking from glass cola bottles for 125 years, but what will we be using in another 125 years? Only time will tell.

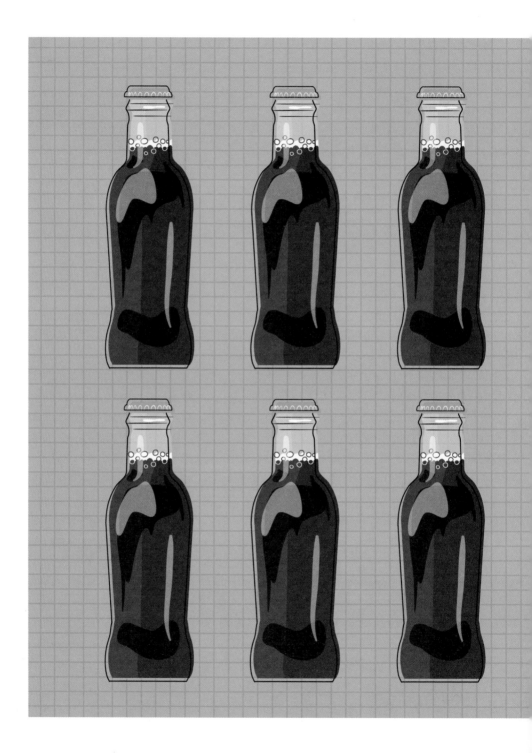

THE
ANSWERS

ANSWERS

CHAPTER 1
STATES OF MATTER

P. 20 IN THE KITCHEN
1. c
2. d
3. a
4. b
5. d

P. 21 SUBSTANCES AND STATES
1. c
2. a
3. e
4. b
5. f
6. d

P. 36 TESTING TEMPERATURES
1. a

2. c Ice is less dense than water.

3. False. Different substances change phase at different temperatures, and so carbon dioxide is a gas at room temperature while water is a liquid.

4. b

P. 37 HEATING AND COOLING
1. d
2. c

3. Starting with the most dense: water, vegetable oil, ice, air.

4. Volume decreases, pressure increases, density increases, and the space between the particles decreases.

5. a
6. c
7. d
8. a
9. b

CHAPTER 2 SOLUTION

P. 54 MOLES
1. $M_r = (12 \times 12) + (22 \times 1) + (11 \times 16) = 342$

2. Moles = 11/342 = 0.03 moles (mol), or 30 mmol.

3. The water has more molecules.
First, turn mass into moles: 100/342 = 0.29 mol for sucrose, and 100/18 = 5.56 mol for water. Now calculate the number of molecules by multiplying the moles by Avogadro's constant.

Sucrose = $0.29 \times 6.02 \times 10^{23} = 1.75 \times 10^{23}$
Water = $5.56 \times 6.02 \times 10^{23} = 33.47 \times 10^{23}$

4. Yes, she could: 61 km (38 miles) deep.

The US has an area of 9.83×10^{16} cm².

It would take 9.83×10^{16} cubes to cover the US, and you have 6.02×10^{23} cubes (one mole). Therefore they could cover the US multiple times.

P. 55 ACIDS AND BASES
1. b
2. b
3. d
4. c
5. b
6. c
7. d (It could be acidic, neutral, or basic.)
8. a

P. 60 BALANCING EQUATIONS
1. Because each water molecule comprises two hydrogen atoms and one oxygen atom, all the hydrogen atoms would get used up reacting with just half of the oxygen atoms, as each mole contains the same number of atoms. This would produce half a mole of water and leave behind half a mole of unused oxygen atoms.

2. a

3. The reaction produces water. The missing molecule is H_2O.

4. $2Au (s) + 3Cl_2 (g) \rightarrow AuCl_3 (s)$

5. a $TiCl_4 + 2H_2O \rightarrow TiO_2 + 4HCl$
Two H_2O molecules are required to provide four hydrogen atoms for the reaction with the four chlorine atoms in $TiCl_4$, and two oxygen atoms are required for the reaction with the single titanium atom.

b $4Fe + 3O_2 \rightarrow 2Fe_2O_3$
Both sides need to have the same number of oxygen atoms as well as the same number of iron atoms. Iron is easy—just start off with 2Fe on the left. Oxygen needs a number of atoms that is divisible by two (for the left side) and a number divisible by three (for the right side). The lowest number divisible by both two and three is six, which leaves six oxygen atoms on each side. This means the reaction begins with three molecules of O_2 and produces two molecules of Fe_2O_3.

c $C_{12}H_{22}O_{11} + 12O_2 \rightarrow 12CO_2 + 11H_2O$
Twelve molecules of molecular oxygen are required to react with the 12 carbon atoms in $C_{12}H_{22}O_{11}$ to produce 12 molecules of CO_2. This leaves the 22 hydrogen atoms and 11 oxygen atoms to react together to produce 11 water molecules.

6. Hydrochloric acid + sodium hydroxide $\rightarrow$ sodium chloride + water.

7. d The products of the reaction are detailed in the right-hand side of a balanced equation; the state of each of the reactants is shown in parentheses; and the molar ratio is shown by the number of atoms and molecules in the equation. But a balanced equation doesn't contain any information about the rate of a reaction, or how fast it happens.

8. False. Reactions can have many different products; even one starting material could decompose into multiple products.

P. 61 THE LAB
1. d
2. f
3. a
4. i
5. g
6. b
7. h
8. c
9. e

CHAPTER 3 CHEMICAL MAKEUP

P. 76 PERIODIC SUCCESS
1. d
2. f
3. a
4. h
5. g
6. c
7. b
8. e

P. 77 ELECTRONS
1. Element number 23 has 23 electrons.
Gold is element number 79 and has 79 electrons.

2. a Oxygen has six electrons in its valence shell. It has eight electrons in total, but the first two are closer to the nucleus and not part of the valence shell.

b Oxygen needs two more electrons to fill its valence shell.

3. **a** Argon has eight electrons in its valence shell.

b Argon can fit eight electrons in its valence shell.

c Argon does not want to gain or lose any electrons because its shell is full; it therefore does not readily form any bonds.

4. **a** Sodium has one electron in its valence shell and wants to get rid of it, reacting violently when it does so. This electron loss happens quickly and easily, meaning sodium is a reactive metal.

b Sodium loses its outer electron and becomes a positive ion.

c Sodium's outer electron is acquired by the chlorine. This leaves sodium positively charged and chlorine negatively charged.

5. b

6. Covalent (carbon has four electrons in its valence shell and prefers to share electrons rather than steal or get rid of them).

P. 83 CHALLENGE 1

1. Dimethyl ether.

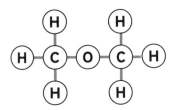

2. Methane.

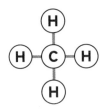

3.

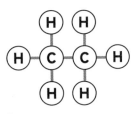

No.

4. Pentane has three structural isomers.

n-pentane.

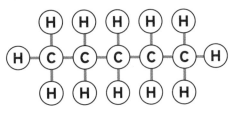

Isopentane.

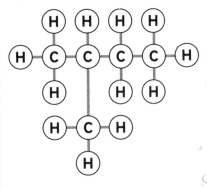

Neo-pentane.

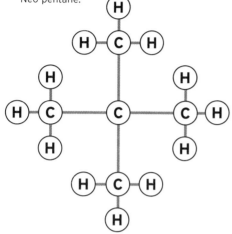

CHALLENGE 2:

Carbon dioxide.

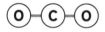

P. 84 VOLUMES AND SURFACES AREAS

1. $2\pi rh$ represents the sides of the cylinder and $2\pi r^2$ represents the circular ends.

2. **a** The radius is 3.09 cm.
   b The surface area is 273 cm².

3. **a** The cube is 15,625 cm³.
   b The surface area of the cube is 3,750 cm².
   c The volume of the hole is 491 cm³.
   Therefore, the new volume of the cube is 15,625 - 491, which equals 15,134 cm³.

4. There are many household examples of objects with large surface areas relative to their size. A good example is a sponge—the deep pores in the structure allow water to be stored. Paper towels and tissues have similar structures. Diapers also comprise materials with large surface areas to lock away moisture quickly.

CHAPTER 5 LOOKING CLOSELY AT COLA

P. 130 COLA

1. c
2. d
3. b
4. a
5. d
6. b
7. d
8. b
9. a
10. **b** The high pressure would allow the gas to stay dissolved in the liquid and it wouldn't try to escape.
11. b
12. **c** They are both made of calcium carbonate.
13. d

CHAPTER 6 COLA ON THE GO

P. 146 POLYMERS

1. c
2. d
3. b
4. d
5. a
6. b
7. c
8. b
9. d
10. d

P. 147 RECYCLING

1. c
2. d
3. b
4. c
5. b
6. c
7. b
8. d

INDEX

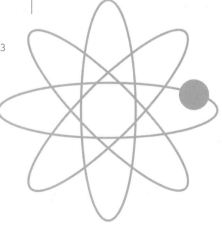

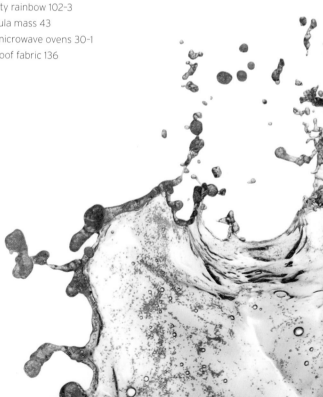

PICTURE CREDITS

SHUTTERSTOCK

4, 70, 71: © SATJA2506

5, 40, 97: © Nito

5, 52, 53, 130: © Love the wind

7, 33, 36: © Triff

10-11: © Natali Zakharova

11, 159: © Handmade Pictures

12, 13, 20, 62, 63, 122, 126, 127, 154, 155: © Alena Ohneva

15: © Melica

16, 17: © Nataly Studio

18: © Passakorn Umpornmaha

21, 25: © Alexey Kljatov

22, 23, 32, 33: © Mariyana M

26: © E.O.

28, 29: © dkidpix

30 top: © bsd

32 bottom: © Kozini

43, 54: © M. Unal Ozmen

44, 45, 159: © Mariyana M

45: © Mrcmos

46: © Hong Vo

50: © Kaiskynet Studio

55 top: © Irin-K

55: © Protasov AN

56 top: © Duda Vasilii

66: © Cigdem

68-69: © Humdan

70 middle: © Ang Intaravichian

72: © Djomas

73: © Oleg GawriloFF

77: ©.Macro Wildlife

78, 79: Studio Vin

80 top: © Apisit Burawannint

80 bottom: © Stockphotograf

81 top: © Baibaz

81 bottom: © Kubais

83: © AJT

86 top, 87: © Schtiel

88-89: © Narudon Atsawalarpsakun

92: © Jiri Hera

93 bottom: © Jiri Hera

96: © Stock Image

97 left: © Taiga

98-99: © Jacek Chabraszewski

100 top: © Yeti Studio

100: © Evgeny Karandaev

104 top, 114: © Ydumortier

105 left: © Danny Smythe

105 right: © Anna Garmashevska

106: © Michael Puche

107: © Goodluz

108 right: © Hekla

110: © Vandame

111 top: © Matveev Aleksandr

111 bottom: © Ollyy

112 bottom left: © Albert Russ

118, 119: © Gyvafoto

122: © Adam Gilchrist

124, 125: © Kirill Aleksandrovich

127 bottom: © Speedkingz

129: © Lightspring

134: © Duxx

135 top: © Africa Studio

136: © Kitch Bain

137: © Pavelis

139 top: © Ombra Estudi

139 bottom: © Fedorova Nataliia

140, 141, 146, 147: © Winai Tepsuttinun

142: © RHJ Phtoto and illustration

143: © Fabrika Simf

144 top, 145, 146: © Dan Kosmayer

144 bottom: © Totem Art

145 top: © Alba Alioth

148-149 (top): Spixel

148-149 (bottom): © Normana Karia

152, 153: © Alena Ohneva

157 top: © Natali Zakharova

Unless otherwise stated, illustrations are by Rob Brandt. Every effort has been made to credit the copyright holders of the images used in this book. We apologize for any unintentional omissions or errors and will insert the appropriate acknowledgment to any companies or individuals in subsequent editions of the work.